TRANSLATIONS OF MATHEMATICAL MONOGRAPHS

VOLUME 62

Kleinian Groups and Uniformization in Examples and Problems

by S. L. KRUSHKAL
B. N. APANASOV
N. A. GUSEVSKIĬ

American Mathematical Society · Providence · Rhode Island

КЛЕЙНОВЫ ГРУППЫ И УНИФОРМИЗАЦИЯ
В ПРИМЕРАХ И ЗАДАЧАХ

С. Л. КРУЩКАЛЬ, Б. Н. АПАНАСОВ И Н. А. ГУСЕВСКИЙ

«НАУКА», НОВОСИБИРСК, 1981

Translated from the Russian by H. H. McFaden
Translation edited by Bernard Maskit

1980 *Mathematics Subject Classification* (1985 *Revision*). Primary 30-01, 30-02, 20H10, 30F10, 30F35, 30F40; Secondary 30C25, 30C60, 30F20, 32G15, 57M12, 58H15.

ABSTRACT. This monograph gives a unified exposition of all the main areas and methods of the theory of Kleinian groups and the theory of uniformization of manifolds. It features a large number of examples, problems, and unsolved problems, many presented for the first time. This is the first publication of this kind in the literature. The book is intended for researchers, graduate students, and undergraduates. It permits nonspecialists to quickly acquaint themselves with contemporary problems in this area of mathematics.
Bibliography: 310 titles. Figures: 67.

Library of Congress Cataloging-in-Publication Data

Krushkal′, S. L. (Samuil Leĭbovich)
Kleinian groups and uniformization in examples and problems.
(Translations of mathematical monographs; v. 62)
Translation of: Kleĭnovy gruppy i uniformizat͡sii͡a v primerakh i zadachakh.
Bibliography: p.
Includes index.
1. Kleinian groups. I. Apanasov, B. N. (Boris Nikolaevich) II. Gusevskiĭ, N. A. (Nikolaĭ Aleksandrovich) III. Title. IV. Series.
QA331.K78513 1986 516.3′73 85-15058
ISBN 0-8218-4516-0

Contents

Foreword

The theory of Kleinian groups was established as far back as the end of the last century, mainly in work of Klein and Poincaré. This theory was created and developed due to the needs of various areas of mathematics: the theory of differential equations, function theory, geometry, topology, number theory, and so on. The theory of Kleinian groups and functions automorphic with respect to them is now one of the most beautiful areas of mathematics and has progressed very far. One of the stimuli promoting the evolution of this theory was the classical problem of uniformization of Riemann surfaces and multivalent analytic functions, solved with different methods by Klein, Poincaré, and Koebe. This has been a source for the development of many fruitful ideas and methods in mathematics.

In the past 20 years the theory of general Kleinian groups and the corresponding theory of uniformization have experienced a rejuvenation. The reawakened interest in these questions was due, in particular, to the development of new powerful methods in topology, the theory of functions of several complex variables, and the theory of quasiconformal mappings.

The number of papers in this area is constantly growing, and it is difficult to imagine a book that could completely reflect the modern state of the theory. Although there are good expositions of its foundations in the well-known monographs by Fricke and Klein, Ford, Appell and Goursat, Lehner, Nevanlinna, and others, none of them is entirely satifactory for the modern reader.

Our set purpose is to give the reader the main contours of this extensive area of research, with special emphasis on examples and problems (which, as a rule, are passed over in the established forms of exposition). They are of varying degrees of difficulty and, in our view, clear up nicely some subtle points of the theory. Most of the examples are also of independent interest.

We count on an active and critical reader, and as food for reflection we provide a large number of exercises, problems, and even unsolved problems. The exercises bear an academic character on the whole, but the problems present more essential results not mentioned in the general theory, with the original sources indicated. The large number of examples and problems enable us to quickly introduce the reader, not only to the actual state of the contemporary theory, but also to the problems and methods which the theory uses.

The main part of the material presented is also contained in numerous journal articles. Some of the material, especially the examples, is presented for the first time. The same applies to many of the exercises. The book was conceived from various special courses and seminar discussions at the Institute of Mathematics of the Siberian Branch of the Academy of Sciences of the USSR and at Novosibirsk State University.

We hope that this book will enable beginning mathematicians to rapidly enter the contemporary circle of problems, and at the same time will be useful both to specialists in the area and to mathematicians working in many other areas.

The authors
Akademgorodok,
Novosibirsk,
November 1979

Editor's Preface to the English Translation

This most unusual book is both a list of basic facts concerning Kleinian groups and uniformization, and a comprehensive compendium of the primary literature in these and related topics. We expect it to be a very valuable resource—as a list of important results, many of which are not elsewhere in print, as a general guide to the literature, and as a specific guide to the Russian literature, which is not as well known here as it deserves to be. The mathematical community owes a debt of gratitude to the three authors.

The book covers an enormous amount of material; while the statements are in general clearly written, almost all of them come without proof. The authors provide references to primary sources for most of the material; however some of their statements are unsupported.

The editor has tried to find the appropriate balance between respecting the flavor of the original, and seeing to it (as best he can) that the book not only is correct, but also does not mislead the unwary. While the editor has tried to make appropriate comments where he found incorrect or misleading statements, he cannot guarantee that he has found them all, especially since some of the errors or gaps may occur in the original literature. Clearly, the specialist should consult the primary sources, with the usual caution.

CHAPTER I

General Properties of Kleinian Groups

§1. Main concepts

1. ***Main definitions.*** Let $\mathcal{M}_2$ be the group of all linear fractional mappings

$$\gamma(z) = (az+b)/(cz+d), ad-bc = 1 \quad (a,b,c,d \in \mathbf{C}) \tag{1}$$

of the extended complex plane $\overline{\mathbf{C}} = \mathbf{C} \cup \{\infty\}$; this group is called the *Möbius group*, and is a simple three-dimensional complex Lie group. There is a natural isomorphism

$$\mathcal{M}_2 \cong SL(2,\mathbf{C})/\{\pm I\},$$

where I is the 2×2 identity matrix.

Recall that a transformation $\gamma \neq I$ is said to be *elliptic* if the square $\operatorname{tr}^2 \gamma = (a+d)^2$ of its trace satisfies the inequality $0 \leq \operatorname{tr}^2 \gamma < 4$, *hyperbolic* if $\operatorname{tr}^2 \gamma > 4$, *parabolic* if $\operatorname{tr}^2 \gamma = 4$, and *loxodromic* if $\operatorname{tr}^2 \gamma \in \mathbf{C} \backslash [0,4]$. The classification of these transformations can also be based, for example, on the properties of their fixed points, the number of which is 1 for parabolic transformations and 2 in the other cases.

We shall consider *discrete* subgroups $\Gamma \subset \mathcal{M}_2$, i.e., subgroups in which the identity is an isolated element.

The group Γ is said to act *discontinuously* at a point $z \in \overline{\mathbf{C}}$ if the stabilizer $\Gamma_z = \{\gamma \in \Gamma \colon \gamma(z) = z\}$ of z in Γ is finite, and z has a neighborhood U_z such that $\gamma(U_z) \cap U_z = \varnothing$ for all $\gamma \in \Gamma \backslash \Gamma_z$ and $\gamma(U_z) = U_z$ for $\gamma \in \Gamma_z$. The set $\Omega(\Gamma)$ of points $z \in \overline{\mathbf{C}}$ at which Γ acts discontinuously is called the *set of discontinuity* of the group Γ. This set is open. Its complement $\Lambda(\Gamma) = \overline{\mathbf{C}} \backslash \Omega(\Gamma)$ is called the *limit set* of the group Γ, and is the set of accumulation points of the *orbits* $\Gamma z_0 = \{\gamma(z_0)\}$, $\gamma \in \Gamma$, for all points $z_0 \in \overline{\mathbf{C}}$.(1)

The group Γ is said to be *(properly) discontinuous* if $\Omega(\Gamma)$ is not empty. Then the limit set $\Lambda(\Gamma)$ is nowhere dense in $\overline{\mathbf{C}}$ and coincides with the closure of the set of fixed points of the nonelliptic elements of Γ. The stabilizer Γ_z is trivial for the points $z \in \Omega(\Gamma)$ that are not fixed points of elliptic elements in Γ, but the stabilizer of each elliptic fixed point of Γ (and only of such points) is

(1) *Translation editor's note*: For an elementary group, there can be an exceptional point.

a cyclic group of finite order. It is clear that a discontinuous group is discrete; the converse is not true in general (see Example 15). It can be shown that the limit set $\Lambda(\Gamma)$ of a discontinuous group is empty, or consists of one or two points, or is infinite. If $\operatorname{card} \Lambda(\Gamma) \leq 2$, then Γ is said to be *elementary*. Such groups can be completely described (see Exercises 8 and 12).

A discontinuous group for which $\Lambda(\Gamma)$ consists of more than two points is called a *Kleinian group*. For each such group $\Lambda(\Gamma)$ is a *perfect* set of positive logarithmic capacity. Sometimes the elementary groups are also put among the Kleinian groups.

The quotient $\Omega(\Gamma)/\Gamma$ is a two-dimensional manifold with a natural complex (conformal) structure in which the projection $\pi\colon \Omega(\Gamma) \to \Omega(\Gamma)/\Gamma$ is holomorphic. It is a finite or countable union $\bigcup_j S_j$ of Riemann surfaces S_j (these surfaces are said to represent the group Γ). The mapping π is a covering which is branched over the projections of the points $z \in \Omega(\Gamma)$ with nontrivial stabilizers Γ_z, i.e., the elliptic fixed points, and the branching order $\nu(p)$ at a point $p = \pi(z)$ is equal to the order of the stabilizer Γ_z.

The set $\Omega(\Gamma)$ of discontinuity itself decomposes into connected components Ω_j, whose number is 1, 2 or ∞ (they are also called the *components of the group* Γ itself). The stabilizers

$$\Gamma_{\Omega_j} = \{\gamma \in \Gamma\colon \gamma(\Omega_j) = \Omega_j\}$$

of these components are subgroups of Γ and are themselves Kleinian groups. If Γ_{Ω_j} coincides with Γ, then the component Ω_j is said to be *invariant*. There can be at most two invariant components. If Ω_0 is an invariant component of Γ, then all the remaining components of Γ must be simply connected. Kleinian groups with invariant components have acquired the name *function groups*. Two components Ω_1 and Ω_2 are said to be *equivalent* or *conjugate* if their stabilizers Γ_{Ω_1} and Γ_{Ω_2} are conjugate in Γ; then $\Omega_1 = \gamma(\Omega_2)$ for some $\gamma \in \Gamma$. If $\Omega_1, \Omega_2, \dots$ is a complete collection of mutually nonequivalent components, then

$$\Omega(\Gamma)/\Gamma = \Omega_1/\Gamma_{\Omega_1} \cup \Omega_2/\Gamma_{\Omega_2} \cup \dots. \tag{2}$$

The group Γ is said to have *finite type* over the component Ω_0 if $S_0 = \Omega_0/\Gamma_{\Omega_0}$ is a *Riemann surface of finite type*, i.e., if S_0 is homeomorphic to a closed surface with finitely many deleted points, and Ω_0 is branched over S_0 at finitely many points with branching of finite order. Suppose that the genus of S_0 is p and the number of deleted points q'_j is n_0. Denote by $\nu(q_j) = \nu_j$, $1 < \nu_j < \infty$, the branching order at the existing branch points $q_j \in S_0$, $j = 1, \dots, n_1$, and let $\nu(q'_j) = \infty, n_0 + n_1 = n$. The vector

$$(p, n; \nu_1, \dots, \nu_n) \tag{3}$$

is called the *signature* of the surface S_0 (and also of the corresponding group Γ_{Ω_0} in Ω_0). The group Γ is called a *group of finite type* if the right-hand side of the decomposition (2) contains only finitely many surfaces, and they are all of finite type.

2. *Some classes of Kleinian groups.* a) *Fuchsian groups.* A Kleinian group Γ is said to be *Fuchsian* if it leaves invariant some circle (line) $L \subset \overline{\mathbf{C}}$ with preservation of the direction of circuit. Then $\Lambda(\Gamma) \subset L$. If $\Lambda(\Gamma) = L$, then Γ is called a group *of the first kind*, and a group *of the second kind* if $L \backslash \Lambda(\Gamma) \neq \varnothing$ (in this case $\Lambda(\Gamma)$ is a nowhere dense subset of L). A Fuchsian group of the first kind has two invariant components, while for groups of the second kind $\Omega(\Gamma)$ is a region coinciding with its single component.

A (nonelementary) Kleinian group is Fuchsian if and only if it does not contain loxodromic elements.

A Fuchsian group Γ represents a single Riemann surface (more precisely, two surfaces which are mirror images of each other). On the other hand, the classical Klein-Poincaré uniformization theorem asserts that, *with the exception of some elementary cases* (when the surface S is conformally equivalent to a sphere, a punctured or twice punctured sphere, or a closed surface of genus 1) *every Riemann surface with given branching orders can be uniformized by a Fuchsian group* Γ acting, for example, in the upper half-plane $H = \{z \in \mathbf{C}\colon \operatorname{Im} z > 0\}$, i.e., to within conformality it can be represented in the form H/Γ. If we introduce the *Poincaré hyperbolic metric*

$$ds^2 = |dz|^2/(\operatorname{Im} z)^2 \tag{4}$$

in H, then the elements of Γ become non-Euclidean (hyperbolic) motions of H.

By analogy with the Fuchsian groups, Poincaré proposed for an arbitrary Kleinian group G a similar interpretation based on extending the action of G to the upper half-space $\mathbf{R}^3_+ = \{(x, y, t) \in \mathbf{R}^3\colon z = x + iy = \mathbf{C},\ t > 0\}$ by means of inversions with respect to the corresponding hemispheres centered on the plane $\mathbf{C}$. We shall discuss the details of this in §2, but for the present we note only that the extension can be interpreted also in another way by using quaternions (see [59]).(2) Namely, by identifying a complex number $z = x + iy \in \mathbf{C}$ with the quaternion $x + iy + j0 + k0$ and a point $(x, y, t) \in \mathbf{R}^3_+$ with the quaternion $z + jt = x + iy + jt + k0$, it is possible to define the action of a unimodular matrix $\left(\begin{smallmatrix} a & b \\ c & d \end{smallmatrix}\right) \in SL(2, \mathbf{C})$ in the half-space $\mathbf{R}^3_+$ according to the rule

$$(z + jt) \to (z' + jt') = \lfloor a(z + jt) + b \rfloor \lfloor c(z + jt) + d \rfloor^{-1}.$$

The group G so extended acts discontinuously in $\mathbf{R}^3_+$, and its elements become non-Euclidean motions of $\mathbf{R}^3_+$ if, by analogy with (4), we introduce there the Poincaré metric

$$ds^2 = (dx^2 + dy^2 + dt^2)/t^2 \tag{5}$$

(2) *Translation editor's note*: This reference appears to be incorrect. A correct reference is: A. F. Beardon, *The geometry of discrete groups*, Springer, 1983.

(and in the large the Möbius group $\mathcal{M}_2$ gives the group of all non-Euclidean motions of the hyperbolic space $\mathbf{R}^3_+$).

The Fuchsian groups are (to within conjugation in $\mathcal{M}_2$) discrete subgroups of the real Möbius group $SL(2,\mathbf{R})/\{\pm I\}$; some of their properties bear a purely algebraic character and admit study by algebraic methods.

b) *Quasi-Fuchsian groups.* These groups are a direct generalization of Fuchsian groups. A *quasi-Fuchsian group* is defined to be a Kleinian group Γ leaving invariant some oriented Jordan curve $L \subset \overline{\mathbf{C}}$. Then the limit set $\Lambda(\Gamma)$ is in L. Again, Γ is said to be *of the first* or *second kind* if $\Lambda(\Gamma) = L$ or $L\backslash\Omega(\Gamma) \neq \varnothing$, respectively. The invariant curve L of a quasi-Fuchsian group of the first kind is smooth if and only if this group is Fuchsian. Otherwise, L automatically does not have a tangent on a dense set of fixed points of loxodromic elements of Γ.

The simplest way to construct quasi-Fuchsian groups is as follows. let $C_1, \dots, C_n$ be circles in $\overline{\mathbf{C}}$ arranged so that each of them is tangent from the outside to the two neighboring ones and does not intersect the others. Consider the group generated by the inversions with respect to these circles, and take its subgroup Γ_0 of index 2 consisting of superpositions of an even number of inversions. This subgroup is quasi-Fuchsian; its invariant curve lies in $\overline{\bigcup_k \operatorname{int} C_k}$ and passes through the points of tangency. Another, very universal, method of obtaining quasi-Fuchsian groups is based on the use of quasiconformal mappings (see §4).

A quasi-Fuchsian group Γ uniformizes two Riemann surfaces D_1/Γ and D_2/Γ, where D_1 is the interior and D_2 the exterior of the invariant curve L; and these surfaces are homeomorphic. Moreover, by a theorem of Bers on simultaneous uniformization, *any two* homeomorphic Riemann surfaces of finite type can be made uniform by a *single* quasi-Fuchsian group (see Chapter II).

Finitely generated quasi-Fuchsian groups have several other characteristic properties. Namely, it follows from results of Accola [9] and Ahlfors [15] that if a finitely generated group has two invariant components, then it is necessarily quasi-Fuchsian; however, for infinitely generated groups this is not true, as shown by Example 20. Further, as established by Maskit [180], finitely generated quasi-Fuchsian groups are quasiconformal deformations of Fuchsian groups, i.e., they can be obtained from Fuchsian groups by conjugation by quasiconformal automorphisms of the plane. As will be clear from Examples 36 and 37, infinitely generated quasi-Fuchsian groups need no longer have these properties.

The indicated properties of finitely generated quasi-Fuchsian groups imply, in particular, a strengthening of the property of noninvariant components of function groups mentioned in subsection 1. Namely, if a function group G is finitely generated, then the stabilizers of all its noninvariant components are quasi-Fuchsian groups (as finitely generated groups with two invariant components!), and these components themselves are bounded by quasicircles. Example 20 shows that this is not true for infinitely generated function groups.

c) *Schottky groups.* Each such group is a Kleinian group Γ with generators $\gamma_1, \ldots, \gamma_p$, $p \geq 1$, such that there exist $2p$ disjoint Jordan curves $l_1, l'_1, \ldots, l_p, l'_p$ bounding a $2p$-connected region D for which $\gamma_j(D) \cap D = \varnothing$ and $\gamma_j(l_j) = l'_j$, $j = 1, \ldots, p$. It turns out that here Γ is necessarily free and purely loxodromic, i.e., all its elements $\gamma \in \Gamma \backslash \{I\}$ are loxodromic or hyperbolic. The factor $\Omega(\Gamma)/\Gamma$ is a closed surface of genus p. Myrberg [197] observed that every nonelementary Kleinian group contains a noncyclic Schottky subgroup.

The Koebe uniformization theorem, which will be presented in the second chapter, asserts that *all closed Riemann surfaces can be uniformized by Schottky groups*; for an appropriate choice of a canonical system of generators $a_1, b_1, \ldots, a_p, b_p$ of the fundamental group $\pi_1(\Omega(\Gamma)/\Gamma)$, the covering π here corresponds to the smallest normal subgroup containing $a_1, \ldots, a_p$ (see, for example, [130]).

d) We take Fuchsian groups $\Gamma_1, \ldots, \Gamma_n$ acting in respective disks $U_1, \ldots, U_n$ sufficiently distant from one another and representing compact surfaces U_j/Γ_j of genus p_j. Then by Klein's combination theorem (presented in §3), the group $\Gamma = \langle \Gamma_1, \ldots, \Gamma_n \rangle$ generated by them is a Kleinian and even a function group, and represents $n+1$ surfaces of genus $p_1, \ldots, p_n$, and $p_1 + \cdots + p_n$, respectively (here we can also consider the somewhat more general case when the Γ_j have arbitrary signatures (3)).

e) *Degenerate groups.* These are nonelementary finitely generated Kleinian groups such that the sets of discontinuity are simply connected regions. There is a nonconstructive and very complicated proof of the existence of such groups due to Bers [51], but explicit examples of them have not been constructed until now (see Example 45). We shall discuss these groups in more detail in §5.

3. *Fundamental regions.* The geometric approach to the investigation of Kleinian groups is based on the concept of a fundamental region. A *fundamental region* of a discontinuous group Γ is defined to be a set $F \subset \Omega(\Gamma)$ containing one point from each orbit $\Gamma z_0, z_0 \in \Omega(\Gamma)$, and such that each nonempty component $F \cap \Omega_j$ of it is connected.(3) This set is not open and, in general, not connected. For example, for Schottky groups, F can be taken to be the region D considered in subsection 2, with the points of the curves $l_1, \ldots, l_p$ adjoined to it. The interior of F alone is often called a fundamental region.

The concept of the isometric circle of a linear fractional transformation turns out to be very useful in the study of Kleinian groups and especially in

(3) *Translation editor's note*: This sentence is a literal rendering of the original. Perhaps the last phrase should read "and such that, for each nonempty component $\Omega_j, F \cap \Omega_j$ is connected."

the construction of a fundamental region. Namely, the *isometric circle* of a transformation $\gamma \in \mathcal{M}_2$ for which ∞ is not a fixed point is defined to be the circle $I(\gamma) = \{z\colon |\gamma'(z)| = 1\}$, or $I(\gamma) = \{z\colon |z + d/c| = 1/|c|\}$, if $\gamma(z)$ has the form (1), i.e., the circle about $\alpha_\gamma = -d/c$ with radius $r_\gamma = 1/|c|$ $(c \neq 0)$. A transformation γ carries its isometric circle $I(\gamma)$ into $I(\gamma^{-1})$, and $\alpha_\gamma = \gamma^{-1}(\infty)$.

Ford [78] showed that *the fundamental region of a Kleinian group* Γ *can be taken to be the exterior of all the isometric circles of its elements, more precisely, the set*

$$P(\Gamma) = \{z \in \Omega(\Gamma)\colon \sup_{\gamma \in \Gamma \setminus \{I\}} |\gamma'(z)| < 1\} = \operatorname{ext} \overline{\bigcup_{\gamma \in \Gamma \setminus \{I\}} (\operatorname{int} I(\gamma))}$$

(with part of the boundary points adjoined). This fundamental region is called the *isometric fundamental region.* The sides of the region $P(\Gamma)$ lying in $\Omega(\Gamma)$ are pairwise equivalent (congruent). The isometric fundamental region of a Fuchsian group has the simplest structure. In this case it is bounded by arcs of circles orthogonal to the invariant circle and consists either of two symmetric components or of a single component, while the mappings connecting its equivalent sides generate the whole group. But in the general case the boundary of the isometric fundamental region can be of a very complicated nature (see Example 18).

In many questions, it is more convenient to deal with other fundamental regions. For example, the so-called *normal fundamental polygons* of Dirichlet are often used for Fuchsian groups. Suppose for definiteness that a Fuchsian group Γ acts in the unit disk $U = \{z \in \mathbf{C}\colon |z| < 1\}$, regarded as the non-Euclidean plane with the Poincaré metric $d(z, z')$ generated by the element

$$ds = 2|dz|/(1 - |z|^2). \tag{6}$$

We take a particular point $z_0 \in U$ that is not a fixed point for elliptic elements of Γ, and consider the set

$$P_{z_0}(\Gamma) = \{z \in U\colon d(z, z_0) \leq d(z, \gamma(z_0)), \gamma \in \Gamma\}. \tag{7}$$

This is a closed convex non-Euclidean polygon generating Γ.

A Fuchsian group Γ of the first kind is finitely generated if and only if $P(\Gamma)$ and $P_z(\Gamma)$ have finite non-Euclidean area, and, consequently, the Riemann surface U/Γ is of finite type; in this case the polygons $P(\Gamma)$ and $P_z(\Gamma)$ have finitely many sides in U and, possibly, finitely many vertices on ∂U that are parabolic fixed points and correspond to deleted points on U/Γ.

For each finitely generated Kleinian group G there exists a fundamental region bounded by finitely many pairwise equivalent smooth arcs.

4. ***The limit set.*** The study of the structure of the limit set $\Lambda(G)$ is one of the difficult problems in the theory of Kleinian groups. Since each nonelementary group G contains Schottky subgroups, this set is "sufficiently massive."

However, the question of its planar measure has not been completely solved. Ahlfors conjectured that it is zero for finitely generated groups, and he proved that this is so if C has in $\mathbf{R}^3_+$ a fundamental polyhedron with finitely many faces (see §2), in which case one says that G is *geometrically finite*; this result was improved somewhat in [44] and [65]. However, for example, degenerate groups are not geometrically finite [87]. For infinitely generated groups the measure of $\Lambda(G)$ can be positive (see Examples 30, 36 and 39).

Very interesting properties are enjoyed by the so-called *residual limit set* $\Lambda_0(G)$, which consists of the points in $\Lambda(G)$ which do not lie on the boundary of any component of $\Omega(G)$. It was studied by Abikoff [4]. In particular, for finitely generated groups $\Lambda_0(G)$ is empty if and only if G has an invariant component or contains a quasi-Fuchsian subgroup of index 2. On this path Abikoff introduced the interesting class of *web groups*; these are the finitely generated Kleinian groups for which the stabilizers of all their components are quasi-Fuchsian groups of the first kind (see Example 23, where there is a construction of such a group that is not quasi-Fuchsian).

Fundamental sets can also be distinguished in the limit set $\Lambda(G)$ (see Example 32). However, Krushkal′ [131] showed that for many groups such fundamental sets turn out to be *nonmeasurable*. Here there arise questions about the possibility of existence of corresponding functions and deformations concentrated only on the limit set (see also [68]). These questions have not been completely answered.

§2. Kleinian groups in space

Until now we have considered groups whose elements are conformal automorphisms of the plane. The transition to the spatial case has its own characteristic features. Let us dwell on them in more detail.

Denote by $\mathcal{M}_n$ the group of all([4]) conformal (Möbius) automorphisms of the extended Euclidean space $\overline{\mathbf{R}}^n = \mathbf{R}^n \cup \{\infty\}$, $n \geq 3$. The definitions of discrete and Kleinian subgroups $G \subset \mathcal{M}_n$, the limit set $\Lambda(G)$, the set $\Omega(G)$ of discontinuity and a fundamental region $F(G)$ are entirely the same in this case as for the plane. The natural projection $\pi\colon \Omega(G) \to \Omega(G)/G$ introduces a conformal structure in $\Omega(G)/G$ (in a way analogous to that in the planar case), which turns it into a union of n-dimensional Riemannian manifolds.([5]) In particular, if $\Omega(G)$ has two invariant components Ω_1 and Ω_2 which are the interior and exterior, respectively, of some ball B^n, then Ω_1/G and Ω_2/G are conformally equivalent complete Riemannian manifolds of constant negative curvature. See §§7 and 8 in Chapter II for more details about the structure of such manifolds (in the case $n \geq 3$).

([4]) *Translation editor's note*: The word "directly" should be inserted here.

([5]) *Translation editor's note*: These need not be manifolds near the projections of fixed points.

Suppose that a mapping $g \in \mathcal{M}_n$ is not a similarity mapping, i.e., $g(\infty) \neq \infty$. Then it is possible to introduce the concept of the isometric sphere of this mapping. This is defined to be the $(n-1)$-dimensional sphere

$$I(g) = \{x = (x_1, \dots, x_n) \in \mathbf{R}^n \colon |g'(x)| = 1\},$$

where $|g'(x)| = |dg(x)|/|dx|$ is the linear dilation at the point x. Here the mapping g itself can be represented as a superposition $U \circ O \circ J$ of an inversion J with respect to the isometric sphere $I(g)$, a reflection O with respect to the hyperplane L such that $I(g^{-1})$ is the mirror image of $I(g)$, and, perhaps, a rotation U about the center of $I(g^{-1})$ (see [26]).

Since each Möbius mapping $g \in \mathcal{M}_n$ is a superposition of finitely many inversions with respect to spheres in $\overline{\mathbf{R}}^n$, it automatically extends to the upper half-plane $\mathbf{R}^{n+1}_+ = \{x \in \mathbf{R}^{n+1} \colon x_{n+1} > 0\}$ if we consider inversions with respect to the corresponding hemispheres in $\overline{\mathbf{R}}^{n+1}_+$ with the same centers; moreover, the extension $\hat{g} \in \mathcal{M}_{n+1}$ of g thus obtained leaves the half-space $\mathbf{R}^{n+1}_+$ invariant. Considering that $\mathbf{R}^{n+1}_+$ is conformally equivalent to the $(n+1)$-dimensional ball, we find that the mapping $\hat{g}$ must have at least one fixed point $p \in \overline{\mathbf{R}}^{n+1}_+$. If the fixed point p lies in $\mathbf{R}^{n+1}_+$, then the mapping $g \in \mathcal{M}_n$ is said to be *elliptic*. If p lies on the boundary hyperplane $\overline{\mathbf{R}}^n$ and is unique, then $g \in \mathcal{M}_n$ is said to be *of parabolic type*. If $\hat{g}$ has two fixed points in $\overline{\mathbf{R}}^n$, then g is said to be *of loxodromic type*. If $\hat{g}$ has three fixed points in $\mathbf{R}^n$, then $\hat{g}$ (and hence also g) leaves invariant the circle C through them, along with the 2-dimensional plane $\mathbf{C}$ containing C. Therefore, $g|_{\mathbf{C}} \equiv I$ (the identity mapping) and, consequently, g is elliptic. A more detailed classification of Möbius mappings $g \in \mathcal{M}_n$ and their properties can be found in [29].

We dwell on the nature of the mappings in these classes. It can be shown that an elliptic mapping g is such that $\hat{g}$ is conjugate to an element in the orthogonal group $O(n+1)$ (and, consequently, g may fail to have fixed points in $\overline{\mathbf{R}}^n$). A mapping g of parabolic type can be conjugate to an element in the group of Euclidean isometries of $\mathbf{R}^n$. A mapping g of loxodromic type is conjugate to a mapping of the form $\gamma(x) = t \cdot A(x)$, where $t \in \mathbf{R}, t \neq 1$ and $A \in O(n)$. If also $t > 0$ and $A = I$, then g is called a *hyperbolic mapping*. The dimension of the set of fixed points of a mapping $g \in \mathcal{M}_n$ is at most $n-2$. In the case of equality the set of fixed points is an $(n-2)$-dimensional sphere, and g is an elliptic mapping.

Useful in studying the group $\mathcal{M}_n$ is the matrix representation of it in the Lorentz group $O(n+1, 1)$; this is based on establishing an isometry $\varphi \colon \mathbf{R}^{n+1}_+ \to \mathbf{H}^{n+1}$ of the two models of $(n+1)$-dimensional hyperbolic space which result from introducing in $\mathbf{R}^{n+1}_+$ and

$$\mathbf{H}^{n+1} = \left\{ (x_0, x_1, \dots, x_{n+1}) \in \mathbf{R}^{n+2} \colon x_0^2 - \sum_{j=1}^{n+1} x_j^2 = 1, x_0 \geq 1 \right\}$$

the respective metrics

$$ds^2 = \sum_{j=1}^{n+1} dx_j^2/x_{n+1}^2, \qquad d_1s^2 = -dx_0^2 + \sum_{j=1}^{n+1} dx_j^2 \tag{8}$$

(cf. (5) and (6)). The representation is obtained as follows. The extensions of Möbius mappings $g \in \mathcal{M}_n$ to $\mathbf{R}_+^{n+1}$ and, correspondingly, the Lorentz transformations on $\mathbf{H}^{n+1}$ are motions in these hyperbolic spaces. Since any mapping $g \in \mathcal{M}_n$ with $g(\infty) \neq \infty$ can be written in the form $g(x) = r \cdot U \circ J(x+y) + y'$ $(y, y' \in \mathbf{R}^n)$, it suffices to establish only the form of the Lorentz transformations corresponding to the following: 1) a translation $x \to x + y$; 2) an orthogonal transformation $x \to U \cdot x$, $U = (u_{ij})_{n \times n}$; 3) an inversion $J(x) = x/|x|^2$; and 4) a dilation $x \to rx$, $r > 0$. It turns out that the following $(n+2, n+2)$-matrices correspond to them:

$$1) \quad \begin{pmatrix} (|y|^2+2)/2 & y_1 \cdots y_n & -|y|^2/2 \\ y_1 & 1 \cdots 0 & -y_1 \\ \vdots & \vdots \quad \vdots & \vdots \\ y_n & 0 \cdots 1 & -y_n \\ |y|^2/2 & y_1 \cdots y_n & (-|y|^2+2)/2 \end{pmatrix};$$

$$2) \quad \begin{pmatrix} 1 & 0 \cdots 0 & 0 \\ 0 & u_{11} \cdots u_{1n} & 0 \\ \vdots & \vdots \quad \vdots & \vdots \\ 0 & u_{n1} \cdots u_{nn} & 0 \\ 0 & 0 \cdots 0 & 1 \end{pmatrix}; \qquad 3) \quad \begin{pmatrix} 1 & 0 \cdots 0 & 0 \\ 0 & 1 \cdots 0 & 0 \\ \vdots & \vdots \quad \vdots & \vdots \\ 0 & 0 \cdots 1 & 0 \\ 0 & 0 \cdots 0 & -1 \end{pmatrix};$$

$$4) \quad \begin{pmatrix} (r^2+1)/2r & 0 \cdots 0 & (r^2-1)/2r \\ 0 & 1 \cdots 0 & 0 \\ \vdots & \vdots \quad \vdots & \vdots \\ 0 & 0 \cdots 1 & 0 \\ (r^2-1)/2r & 0 \cdots 0 & (r^2+1)/2r \end{pmatrix}.$$

This representation is contained, for example, in the article [28] of Apanasov; applications are given in his papers [28], [30], and [31]. The article [245] appeared later and repeats part of these results.

We take a particular point $y \in \overline{\mathbf{R}}^n$ and define for a mapping $g \in \mathcal{M}_n$ a function analogous to the Jacobian $\|g'(x)\|$. For this, consider a $\gamma \in \mathcal{M}_n, \gamma(y) = \infty$, and let $\alpha_y(g, x) = \|(\gamma \circ g \circ \gamma^{-1})'(\gamma x)\|$. The function α_y does not depend on the choice of $\gamma \in \mathcal{M}_n$. See [34] for details on its properties. By using the functions $\alpha_y(g,x)$ it is possible to obtain a fundamental region convenient in applications for a Kleinian group $G \subset \mathcal{M}_n$. To do this, take a point $y \in \Omega(G)$ such that $g(y) \neq y$ for all $g \in G\backslash\{I\}$, and let

$$F_y(G) = \{x \in R^n : \alpha_y(g,x) < 1, g \in G\backslash\{I\}\}. \tag{9}$$

For $y = \infty$ this fundamental polyhedron coincides with the *isometric fundamental polyhedron*

$$P(G) = \left\{ x \in \Omega(G) \colon \sup_{g \in G\backslash\{I\}} |g'(x)| < 1 \right\} = \bigcap_{g \in G\backslash\{I\}} \operatorname{ext} I(g).$$

Moreover, it can be shown that if G leaves the half-space $\mathbf{R}^n_+$ invariant, then the *normal fundamental polyhedron* of Dirichlet about the point $y \in \mathbf{R}^n_+$ defined by

$$P_y(G) = \{x \in \mathbf{R}^n_+ : d(x,y) < d(x, g(y)), g \in G\backslash\{I\}\}$$

coincides with the polyhedron $F_y(G) \cap \mathbf{R}^n_+$; here $d(x,y)$ is the hyperbolic distance in $\mathbf{R}^n_+$, and $F_y(G)$ is given by (9).

A Kleinian group $G \subset \mathcal{M}_n$ is said to be *Fuchsian* in $\mathbf{R}^n$ if there exists a ball B^n (or a half-space) that is invariant under G. Moreover, G is called a Fuchsian group *of the first* (*second*) *kind* if $\Lambda(G) = \partial B^n$ (respectively, $\partial B^n \backslash \Lambda(G) \neq \varnothing$).

A Kleinian group $G \subset \mathcal{M}_n$ is said to be *quasi-Fuchsian* if there exists a Jordan surface $S \subset \overline{\mathbf{R}}^n$ with interior and exterior homeomorphic to an n-dimensional ball and invariant under G.

An important class of Kleinian groups in the half-space $\overline{\mathbf{R}}^n$ is the class of groups having fundamental polyhedra in $\mathbf{R}^{n+1}_+$ bounded by finitely many faces (geometrically finite Kleinian groups). As already noted, for $n \geq 3$ it no longer coincides (unlike Fuchsian groups in the plane, i.e. for $n \geq 2$) with the class of finitely generated groups (see also Examples 46 and 51). A sufficient condition for a discrete group $G \subset \mathcal{M}_n$ to belong to the class of geometrically finite groups is that the hyperbolic volume of a fundmental polyhedron of G in $\mathbf{R}^{n+1}_+$ be finite (see Selberg [298], Garland and Raghunathan [81], and Wielenberg [245]); here the limit set $\Lambda(G)$ of G coincides with $\overline{\mathbf{R}}^n$, i.e., G is not Kleinian in $\overline{\mathbf{R}}^n$. We introduce the following concepts for characterizing the geometrically finite Kleinian groups $G \subset \mathcal{M}_n$. A limit point $z \in \Lambda(G)$ is called a *point of* (*conical*) *approximation* of a group $G \subset \mathcal{M}_n$ if for any hyperbolic line $\sigma \subset \mathbf{R}^{n+1}_+$ ending at z there exists a compact set $K \subset \mathbf{R}^{n+1}_+$ and a sequence of distinct elements $g_m \in G$ such that $g_m(\sigma) \cap K \neq \varnothing$. A point $z \in \overline{\mathbf{R}}^n$ that is a fixed point of a parabolic element of a group $G \subset \mathcal{M}_n$ is called a *parabolic vertex* of G (see [44]) if one of the following conditions holds:

1) The stabilizer $G_z = \{g \in G : g(z) = z\}$ has a free abelian subgroup of rank n.

2) There exists an open subset $U_z \subset \Omega(G) \subset \overline{\mathbf{R}}^n$ such that, for some mapping $h \in \mathcal{M}_n$ with $h(z) = \infty$ and for some $t > 0$,

$$1) \qquad \{x \in \mathbf{R}^n \colon \sum_{i=k+1}^{n} x_i^2 > t\} \subset h(U_z),$$

$$2) \qquad \gamma(U_z) = U_z \quad \text{for all } \gamma \in G_z,$$

$$3) \qquad g(U_z) \cap U_z = \varnothing \quad \text{for all } g \in G \backslash G_z.$$

Here the integer $k, 1 \leq k \leq n-1$, is equal to the rank of a maximal free abelian subgroup of the stabilizer G_z, and the last two conditions give the property of strict invariance of the set U_z with respect to the subgroup G_z of G.

Using these concepts, we can give a criterion for discrete groups $G \subset \mathcal{M}_n$ (including Kleinian groups) to be geometrically finite: *a group G in this class is characterized by the condition that its limit set $\Lambda(G)$ consists of points of approximation and parabolic vertices*; see [44] for $n = 2$ and [253] for $n \geq 3$. The case $n \geq 3$ has its own distinctive features, due, in particular, to the fact that there exist Kleinian groups $G \subset \mathcal{M}_n, n \geq 3$, whose parabolic points $z \in \overline{\mathbf{R}}^n$ do not have horiballs $B_z \subset \mathbf{R}^{n+1}_+$ strictly invariant with respect to the stabilizers G_z (if $z = \infty$, then $B_z = \{x \in \mathbf{R}^{n+1}_+ \colon x_{n+1} > r > 0\}$); see [260].

Along with the above analytic criterion it is possible to give criteria of a geometric and topological nature for geometric finiteness. Namely, *a discrete torsion-free group $G \subset \mathcal{M}_n, n \geq 1$, is geometrically finite if and only if any one of the following properties holds* (see Thurston [234] for $n = 2$, and Apanasov [254], [258] for $n \geq 3$):

1) *For some (for any) $\varepsilon > 0$ the ε-neighborhood $U_\varepsilon(N)$ of a minimal convex retract N of the hyperbolic manifold $\mathbf{R}^{n+1}_+/G$ has finite volume.*

2) *The submanifold $N_{[\varepsilon,\infty)} \subset N \subset \mathbf{R}^{n+1}_+/G$ consisting of the points in a minimal convex retract N through which no nontrivial (in N) loops of length less than ε pass is compact.*

We remark that for discrete groups $G \subset \mathcal{M}_n$ with limit set $\Lambda(G) = \overline{\mathbf{R}}^n$ the first characterization of geometric finiteness coincides with finiteness of the volume for $\mathbf{R}^{n+1}_+/G$, since $N = \mathbf{R}^{n+1}_+/G$ in this case. The second characterization of geometric finiteness is connected with a criterion of Marden [168] for geometric finiteness of planar Kleinian groups in terms of neighborhoods of ends of the manifold $\mathbf{R}^3_+/G$.

§3. Combination theorems

One of the most important ways of constructing new Kleinian groups is to construct these groups from simpler groups already at hand. In this case it is common to consider some family $\{G_j\}$ $(j = 1, 2, \ldots)$ of Kleinian groups along with the group $G = \langle G_1, G_2, \ldots \rangle$ generated by the family. The problem is to find conditions under which G is again Kleinian, and to clarify the algebraic structure of G. Such a problem was apparently first solved by Klein [117], who considered the case when G is the free product of the groups $G_1, G_2, \ldots$. A subsequent and far-reaching development of this method of constructing Kleinian groups, which came to be called *the combination method*, was due to Maskit in [180] and [181]. His constructions actually bear a geometric character and are profoundly connected with uniformization theory.

1. ***Klein's combination theorems*** We first describe Klein's method of construction. All the assertions we give will be illustrated by corresponding examples.

THEOREM 1. *Let $\{G_j\}$, $j = 1, 2, \ldots$, be a family of Kleinian groups such that each element $\gamma \in G_j, \gamma \neq I$, $j = 1, 2, \ldots$, has an isometric sphere, and suppose that the following conditions hold:*

a) *The isometric spheres of G_j lie in the closure of the isometric fundamental region $P(G_k)$ of G_k for every $j \neq k$ $(j, k = 1, 2, \ldots)$.*

b) *The intersection of the domains $P(G_j)$, $j = 1, 2, \ldots$, contains a neighborhood of the point at infinity in $\overline{\mathbf{R}}^n$ $(n \geq 2)$.*

Then the group $G = \langle G_1, G_2, \ldots \rangle$ is Kleinian and is the free product of the groups $G_1, G_2, \ldots$, and $P(G) = \operatorname{int} \bigcap_1^\infty P(G_j)$ is a fundamental region of G.

EXAMPLE 1. Let $C_1, C_1', \ldots, C_m, C_m'$ be spheres in $\mathbf{R}^n$ with disjoint interiors such that C_j and C_j' have equal radii, $j = 1, \ldots, m$. Let γ_j be a Möbius mapping such that C_j is the isometric sphere for γ_j, and C_j' is the isometric sphere for γ_j^{-1}. Consider the group $G = \langle \gamma_1, \ldots, \gamma_m \rangle$. Obviously the conditions in Theorem 1 hold and, consequently, the group G is Kleinian and is the free product of the cyclic groups generated by the mappings $\gamma_1, \ldots, \gamma_m$; the exterior of all the spheres C_j, C_j', $j = 1, \ldots, m$, is the isometric fundamental region of G.

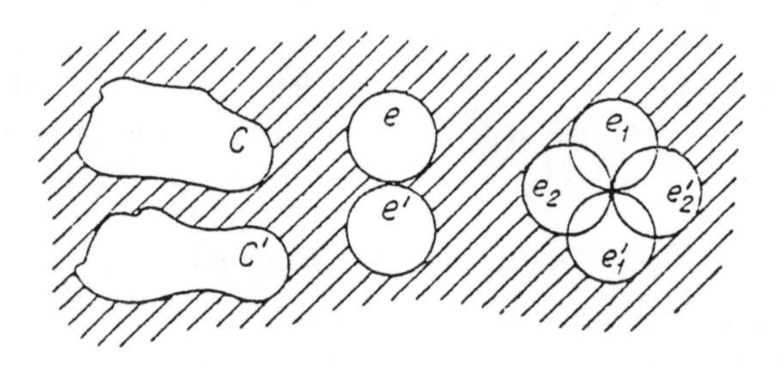

FIGURE 1

THEOREM 2. *Let $G_1, \ldots, G_m$ be Kleinian groups in $\mathcal{M}_n$ with fundamental regions $F_1, \ldots, F_m$. Assume that the exterior F_j is contained in F_k, $j \neq k$, $(j, k = 1, \ldots, m)$. Then the group $G = \langle G_1, \ldots, G_m \rangle$ is Kleinian and is the free product of the groups $G_1, \ldots, G_m$, and $F = \bigcap_1^m F_j$ is a fundamental region of G.*

The class of groups d) in §1 can be taken as an illustrative example. Here is another example.

A Kleinian group $G \subset \mathcal{M}_2$ is called an *extended Schottky group* or a *Schottky group of type* (g, s, m) if it is a free product of g cyclic loxodromic groups, s free abelian groups of rank 2, and m cyclic parabolic groups. An analogous definition can be given in the multi-dimensional case.

EXAMPLE 2. We construct a Schottky group of type $(1, 1, 1)$. Let C and C' be disjoint Jordan curves in $\overline{\mathbf{C}}$ lying outside each other; let e and e' be circles of equal radius tangent to each other from the outside and lying in the exterior of C and C'; let e_1, e_1', e_2, e_2' be circles intersecting at a single point and lying in the exterior of the four previous curves (see Figure 1). Let γ be a linear fractional transformation carrying C into C' such that the interior of C goes into the exterior of C'; let $\tilde{\gamma}$ be a parabolic mapping carrying e into e'; let γ_1 and γ_2 be parabolic mappings carrying e_1 into e_1' and e_2 into e_2', respectively, which generate a free abelian group of rank 2. It is clear that the conditions in Theorem 2 are satisfied and, consequently, the group G generated by the groups $\langle \gamma \rangle, \langle \tilde{\gamma} \rangle$, and $\langle \gamma_1, \gamma_2 \rangle$ are Kleinian. It can be shown that the set $\Omega(G)$ of discontinuity is connected, and $\Omega(G)/G$ is a Riemann surface of genus 2 with two deleted points.

2. *Maskit's combination theorems.* Let G be a Kleinian group on the plane and H a subgroup of it. A topological disk B in $\overline{\mathbf{C}}$ is said to be *precisely invariant under* H if $\gamma(B) = B$ for all $\gamma \in H$, and $\gamma(B) \cap B = \varnothing$ for all $\gamma \in G \backslash H$. If G is of finite type, then H is trivial or cyclic or a finitely generated quasi-Fuchsian group. Since $B \cap \Lambda(G) = \varnothing$, either $J = \partial B$ is a subset of $\Lambda(G)$ of $J \cap \Omega(G)$ is contained in a component of $\Omega(G)$.

THEOREM 1. *Let G_1 and G_2 be Kleinian groups with a common subgroup H, let B_1 and B_2 be disks precisely invariant under H, let $J = \partial B_1 = \partial B_2$, and suppose that the following conditions hold*: a) *H is trivial or cyclic or a quasi-Fuchsian group of the first kind*; b) *$\gamma(B_j) \cap B_j = \varnothing$ for all $\gamma \in G_j \backslash H$, $j = 1, 2$*; c) *if F_j is a fundamental set of G_j, then $F_j' = F_j \cap G_j(B_j) \subset B_j$, $j = 1, 2$; and* d) *$F_j' \cap F_k$ has a nonempty interior.*

Then the group $G = \langle G_1, G_2 \rangle$ is Kleinian and is the free product of the groups G_1 and G_2 with amalgamated subgroup H, and $\Omega(G)/G$ is the union of $(\Omega(G_1) \backslash G_1(B_1))/G_1$ and $(\Omega(G_2) \backslash G_2(B_2))/G_2$, in which $(J \cap \Omega(H))/H \subset (\Omega(G_1) \backslash G_1(B_1))/G_1$ is identified with $(J \cap \Omega(H))/H \subset (\Omega(G_2) \backslash G_2(B_2))/G_2$.

EXAMPLE 3. Let G_1 and G_2 be Fuchsian groups of the first kind with two parabolic generators each and acting in the respective disks U_1 and U_2. We assume that U_1 and U_2 and the fundamental polygons of the groups G_1 and G_2 are as in Figure 2. Here fixed points of parabolic elements of G_1 and G_2 serve as vertices of the fundamental polygons, and the origin of coordinates is a fixed point of a parabolic element h generating a common subgroup H of G_1 and G_2 (the isometric circles of h and h^{-1} are marked by dashes). Let B_1 be the right and B_2 the left half-plane. Then B_1 and B_2 are disks satisfying the conditions of Maskit's theorem and, consequently, the group $G = \langle G_1, G_2 \rangle$ is Kleinian. It can be shown that, in fact, G is a *b-group* containing *accidental parabolic elements*; for example, h is such an element (see §5 and Example 47).

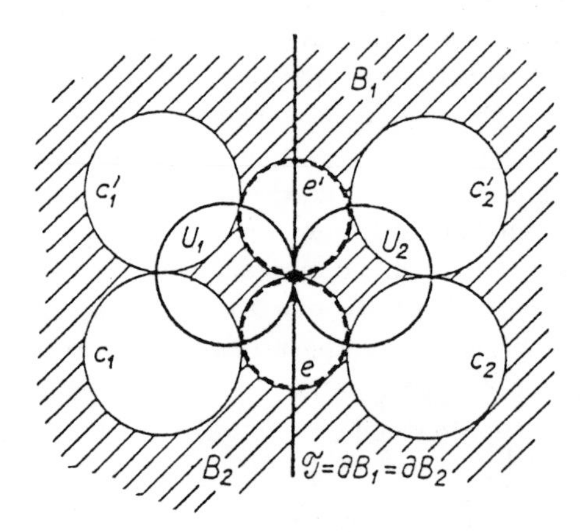

FIGURE 2

THEOREM 2. *Let G_0 be a finitely generated Kleinian group on the plane, H_1 and H_2 subgroups of it, and B_1 and B_2 disks precisely invariant under H_1 and H_2. Assume the following:* a) *H_1 and H_2 are trivial, or cyclic, or finitely generated quasi-Fuchsian groups of the first kind;* b) *there exists a linear fractional mapping $f \notin G_0$ such that $fH_1f^{-1} = H_2$;* c) *$\gamma(B_1) \cap B_2 = \varnothing$ for all $\gamma \in G_0$;* d) *$f(\partial B_1) = \partial B_2$; and* e) *$f(B_1) \cap B_2 = \varnothing$.*

Suppose, further, that G_0 has a fundamental set F with the following properties: f) *$F \cap B_j$ is a fundamental set for the group H_j acting on B_j, $j = 1, 2$;* g) *$F \cap (\overline{\mathbf{C}} \setminus (B_1 \cup B_2))$ has nonzero interior.*

Then the group $G = \langle G_0, f \rangle$ is Kleinian, $\Omega(G)/G$ coincides with the set $(\Omega(G_0) \setminus G_0(B_1 \cup B_2))/G_0$, in which the subsets $\pi(\partial B_1)$ and $\pi(\partial B_2)$ are identified by the mapping $\pi \circ f$ (where $\pi\colon \Omega(G) \to \Omega(G)/G$ is the natural projection), and each relation in G is a consequence of relations in G_0, along with relations arising from the equality $fH_1f^{-1} = H_2$.

EXAMPLE 4. Let C_1 and C_2 be circles of equal radius centered on the real semi-axis, tangent to each other, and lying entirely in the right half-plane. Denote by C_1' and C_2' the symmetric images of C_1 and C_2 with respect to the imaginary axis. Take hyperbolic mappings γ_1 and γ_2 such that C_1 and C_2 are the isometric circles of γ_1 and γ_2, while C_1' and C_2' are the isometric circles of γ_1^{-1} and γ_2^{-1}, respectively. Let G_0 be the group generated by γ_1 and γ_2. By Klein's combination theorem, G_0 is Kleinian and is the free product of the cyclic groups $H_1 = \langle\gamma_1\rangle$ and $H_2 = \langle\gamma_2\rangle$.

Let l_j be the circle about the origin orthogonal to C_j, $j = 1, 2$ (see Figure 3). Consider the mapping $f\colon z \to kz$, where $k = r_2/r_1$ (r_j is the radius of l_j). Denote by B_1 the interior of l_1, and by B_2 the exterior of l_2. Obviously, $f(l_1) = l_2$ and $f(B_1) \cap B_2 = \varnothing$. Since l_1 and l_2 are orthogonal to the isometric circles γ_1 and γ_2, it follows from general properties of linear fractional mappings that B_1 and B_2 are precisely invariant disks with respect to H_1 and H_2, $fH_1f^{-1} = H_2$ and $\gamma(B_1) \cap B_2 = \varnothing$ for all $\gamma \in G_0$. Thus, all the conditions of Theorem 2 hold. Therefore, $G = \langle G_0, f\rangle$ is Kleinian. Moreover, it is not hard to show that G is a Fuchsian group of the first kind.

If instead of the mapping $f(z) = kz$ we consider $f_1(z) = -kz$, then it is not hard to see that the conditions of Theorem 2 again are satisfied and, consequently, the group $G_1 = \langle G_0, f_1\rangle$ is Kleinian. Since f_1 interchanges the upper and lower half-planes, G_1 does not have invariant components. It can be shown that G_1 is a $\mathbf{Z}_2$-extension of a Fuchsian group of the first kind, i.e., contains a quasi-Fuchsian subgroup of index 2.

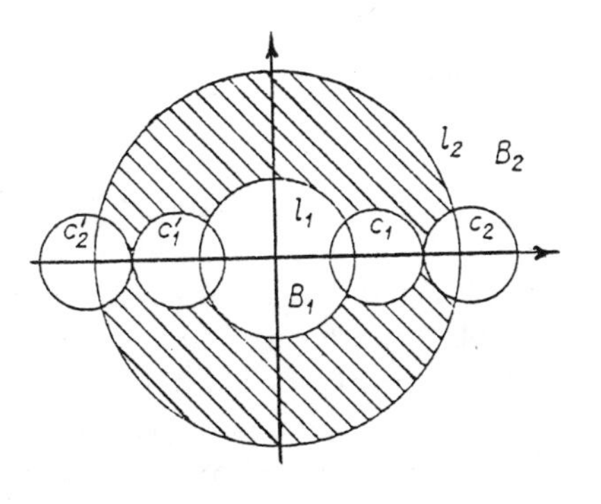

FIGURE 3

EXAMPLE 5. Let G_1 be a Fuchsian group of the first kind leaving invariant the unit circle C. Let $f_t(z)$ be the mapping $z \to tz$, where $t > 0$, and let $G_t = f_t^{-1}G_1f_t$ and $U_t = f_t(U)$. Fix a t_0 large enough that the exterior of a fundamental set of G_1 is contained interior to the part of a fundamental set of G_{t_0} lying in the disk U_{t_0}. Using Klein's combination theorem, we get that the group $G_0 = \langle G_1, G_{t_0} \rangle$ is Kleinian. Let $B_1 = U_1$ and $B_2 = f_{t_0}(U^*)$, where U^* is the region $|z| > 1$, and let $f = f_{t_0}$. It is clear that all the conditions of Theorem 2 (Maskit) are satisfied for G_0 and f; consequently, the group $G = \langle G_0, f \rangle$ is Kleinian.

The combination theorems of Maskit were formulated for planar groups. But (like Klein's theorems) they admit a corresponding generalization to the multi-dimensional case (see Exercise 29).

§4. Quasiconformal mappings and deformations of Kleinian groups

1. ***Quasiconformal mappings on the plane.*** Let D be a region in $\mathbf{C}$. *Quasiconformal mappings* of D are understood to be generalized homeomorphic solutions $w = f(z)$ of Beltrami's equation

$$w_{\bar{z}} - \mu(z) w_z = 0, \tag{10}$$

where $\mu(z)$ is a measurable function in D with

$$\|\mu\|_\infty < 1, \tag{11}$$

and $w_z = \frac{1}{2}(w_x - i w_y)$ and $w_{\bar{z}} = \frac{1}{2}(w_x + i w_y)$ (generalized derivatives) are locally square-integrable. The inequality (11) shows that orientation-preserving homeomorphisms are taken. The mapping f is conformal in the Riemannian metric

$$ds^2 = \lambda(z)|dz + \mu(z)\, d\bar{z}|^2, \qquad \lambda(z) > 0, \tag{12}$$

if it is introduced in D; then angles are preserved almost everywhere. It is convenient to measure the deviation of this mapping from an ordinary conformal mapping by the quantity

$$K(f) = (1 + \|\mu\|_\infty)/(1 - \|\mu\|_\infty), \tag{13}$$

called the *deviation*. It is not hard to verify that $K(f^{-1}) = K(f)$, and the inequality $K(f_1 \circ f_2) \leq K(f_1)K(f_2)$ holds for a superposition $f_1 \circ f_2$. It is clear from (13) that $K(f) \geq 1$, with equality corresponding to the case of conformal mappings (then (10) becomes the Cauchy-Riemann equation $w_{\bar{z}} = 0$).

For $K(f) = K$ the mapping f is also called K-*quasiconformal.* The function $\mu_f(z) = \mu(z)$ is called the *complex characteristic* or *dilation* of f.

Sometimes, following M. A. Lavrent′ev, one considers instead of μ the two real characteristics $p(z) \geq 1$ and $\theta(z)$ connected with μ by the relation

$$\mu(z) = 0 - [(p(z) - 1)/(p(z) + 1)]e^{2i\theta(z)}.$$

They have a simple geometric meaning: the inverse images of infinitesimally small disks in the w-plane are infinitesimally small (characteristic) ellipses in the z-plane with the ratio of the large to the small semi-axes equal to $p(z)$ and the slope of the large semi-axes equal to $\theta(z)$ (if $p(z) = 1$, then $\theta(z)$ is not defined). *The essence of quasiconformality is that in the small such mappings behave* (almost everywhere, in general) *like affine mappings*; indeed,

$$w - w_0 = w_z(z_0)(z - z_0) + w_{\bar{z}}(z_0)(\bar{z} - \bar{z}_0) + o(|z - z_0|) \tag{14}$$

at a point z_0 where $w(z)$ is differentiable.

Quasiconformality can also be defined in an entirely different, purely geometrical, way leading to the same class of mappings. Namely, consider a curvilinear quadrangle, i.e., a Jordan region Q with a distinguished quadruple of boundary points (the vertices) and an ordered pair of opposite boundary arcs. This quadrangle can be mapped conformally with the correspondence of vertices onto a rectangle R, and, if the lengths of the sides of R corresponding to the first and second pairs of sides of Q are denoted by a and b, then the ratio a/b is a conformal invariant. It is called the (*conformal*) *modulus* of the quadrangle Q and denoted by $\operatorname{mod} Q$.

It turns out that *a homeomorphism $f\colon D \to \overline{\mathbf{C}}$ is K-quasiconformal if and only if it changes the modulus of any curvilinear quadrangle $Q \subset D$ by at most a factor of K*:

$$K^{-1}\operatorname{mod} Q \leq \operatorname{mod} f(Q) \leq K \operatorname{mod} Q; \tag{15}$$

here

$$K(f) = \sup_{Q \subset D} (\operatorname{mod} f(Q)/\operatorname{mod} Q), \tag{16}$$

which strengthens the relation (15).

We note the main properties of quasiconformal mappings; proofs of them can be found, for example, in [19], [130], and [151].

1) *For any measurable complex characteristic $\mu(z)$ defined in $\mathbf{C}$ with $\|\mu\|_\infty < 1$ there exists a unique quasiconformal automorphism $f\colon \overline{\mathbf{C}} \to \overline{\mathbf{C}}$ satisfying the Beltrami equation (10) and leaving fixed three specified points, for example,* 0, 1, *and* ∞.

2) *If $\varsigma = f_0(z)$ is a homeomorphic solution of* (10) *in D, then the set of all* (*generalized*) *solutions of this equation in D is exhausted by the formula*

$$f(z) = \Phi[f_0(z)], \tag{17}$$

where Φ is an arbitrary analytic function of ς in the region $f_0(D)$; in other words, any two solutions of the same Beltrami equation can be obtained one from the other by superposition with an analytic function.

Extending μ to $\mathbf{C}\backslash D$ by zero, we get the following facts from (17) and 1):

3) *Any two simply connected regions D and D' of the same conformal type can be mapped one onto the other by a homeomorphic solution of equation* (10) *with an arbitrarily specified μ in D* (*respectively, in D'*), $\|\mu\|_\infty < 1$ (***analogue of the Riemann mapping theorem***).

4) *The conditions of uniqueness for a quasiconformal mapping with specified μ must ensure uniqueness of the analytic function Φ in* (17), *i.e., they bear the same character as in the theory of conformal mappings.*

5) *Quasiconformal mappings of one Jordan region onto another extend to homeomorphisms of the closed regions.*

6) There is a formula for representing quasiconformal automorphisms of the plane that are conformal in a neighborhood of the point at infinity. Namely, *if a measurable function* $\mu(z), \|\mu\|_\infty \leq k < 1$, *in* $\mathbf{C}$ *has compact support, then a solution of* (10) *homeomorphic in* $\mathbf{C}$ *and normalized by the condition* $f(z) = z + O(|z|^{-1})$ *for* $z \to \infty$ *can be represented in the form*

$$f(z) = z - \frac{1}{\pi} \iint_{\mathbf{C}} \frac{\rho(\varsigma)\, d\xi\, d\eta}{\varsigma - z} \equiv z + T\rho(z) \qquad (\varsigma = \xi + i\eta), \tag{18}$$

where the function ρ is in $L_p(\mathbf{C})$ for some $p > 2$ (more precisely, $2 < p < p_0(k)$) and is a solution of the integral equation

$$\rho - \mu \Pi \rho = \mu, \tag{19}$$

in which

$$\Pi\rho = -\frac{1}{\pi} \iint_{\mathbf{C}} \frac{\rho(\varsigma)\, d\xi\, d\eta}{(\varsigma - z)^2} = \frac{\partial T\rho}{\partial z}; \tag{20}$$

here the integral in (20) is understood in the principal value sense. Equation (19) can be solved by the method of successive approximations, and this gives us that $\rho = \mu + \mu\Pi\mu + \mu\Pi(\mu\Pi\mu) + \cdots$.

Formulas (18) and (19) can be used to establish the following important properties of quasiconformal mappings:

7) *The derivatives* f_z *and* $f_{\bar{z}}$ *of a quasiconformal mapping* $D \to \mathbf{C}$ *are locally integrable to some power* $p > 2$ *in* D, *and the mapping is absolutely continuous with respect to planar Lebesgue measure.*

8) *The family of normalized* K*-quasiconformal automorphisms of* $\overline{\mathbf{C}}$ (*or of the disk* U) *is compact.* Moreover, if f^μ denotes the homeomorphic solution of (10) in $\overline{\mathbf{C}}$ leaving fixed the points 0, 1 and ∞ and if $\|\mu\|_\infty \leq k < 1$, then the inequality $|f^\mu(z)| \leq A_1 \|\mu\|_\infty$ holds in any disk $U_R = \{z\colon |z| < R\}$ $(R < \infty)$, and, for any two points $z_1, z_2 \in U_R$,

$$A^{-1}|z_1 - z_2|^{p/(p-2)} \leq |f^\mu(z_1) - f^\mu(z_2)| \leq A|z_1 - z_2|^{(p-2)/p},$$

where the constant A_1 depends only on k and R, while A depends on k, R and $p, 2 < p < p_0(k)$.

9) If $k = \varepsilon$ is sufficiently small, then (18) and (19) give us the following *variation formula for quasiconformal mappings*:

$$f^\mu(z) = z - \frac{z(z-1)}{\pi} \iint_{\mathbf{C}} \frac{\mu(\varsigma)\, d\xi\, d\eta}{\varsigma(\varsigma - 1)(\varsigma - z)} + O(\|\mu\|_\infty^2), \tag{21}$$

with a uniform estimate for the remainder term on compact subsets of $\mathbf{C}$. This implies, in particular, that *if* μ, *as an element of* $L_\infty(\mathbf{C})$, *depends continuously, differentiably, or holomorphically on complex parameters* ω, *then*

$f^\mu(z)$ is a continuous, differentiable, or holomorphic function of ω for each particular z (even in a stronger norm which includes the L_p-norms of the derivatives f_z and $f_{\bar z}$), and the differential of the mapping $\mu \to f^\mu$ at zero is equal to the integral term in (21).

10) *Every quasiconformal automorphism f of the upper half-plane $H = \{z \in \mathbf{C}\colon \operatorname{Im} z > 0\}$ is extended to the whole plane $\mathbf{C}$ by the equality $f(z) = \overline{f(\bar z)}$ for z with $\operatorname{Im} z < 0$; then $\mu_f(z) = \overline{\mu_f(\bar z)}$. This extension also carries H into itself. The automorphisms of H induced by the homeomorphisms $\overline{\mathbf{R}} \to \overline{\mathbf{R}}$ must satisfy a definite condition found by Beurling and Ahlfors* [60].

Closely connected with this condition are properties of the Jordan curves that are images of circles under quasiconformal automorphisms of $\overline{\mathbf{C}}$. Such curves are called *quasicircles*. They may fail to be rectilinear in all their parts, but they have planar measure zero by 7). Ahlfors [19] showed that a closed oriented Jordan curve $L \subset \overline{\mathbf{C}}$ passing through the point at infinity is a quasicircle if and only if, for any three finite points $\varsigma_1, \varsigma_2, \varsigma_3$ on it with ς_2 separating ς_1 and ς_3, the ratio $|\varsigma_2 - \varsigma_1|/|\varsigma_3 - \varsigma_1|$ is bounded by a constant depending only on L (boundedness of the cross-ratio of four points is required for bounded curves); this is equivalent to the condition that the ratio of the diameter of an arc to the length of the chord subtending it is uniformly bounded for L.

In particular, the invariant curves of finitely generated quasi-Fuchsian groups are quasicircles and, consequently, have the properties indicated above.

2. *Quasiconformal mappings of Riemann surfaces and deformations of Kleinian groups.* In the case of mappings of Riemann surfaces the coefficient $\mu(z)$ must also satisfy a certain invariance condition; namely, the form $\mu = \mu(z)\, d\bar z/dz$ must remain invariant under a change of the local parameter z on the given Riemann surface S. Such forms are called *Beltrami differentials.* If local parameters z and z' are connected by a conformal correspondence $z' = h(z)$, then the equality $\mu(z)\, d\bar z/dz = \mu(z')\, d\bar z'/dz'$ gives the correspondence $\mu(h(z))\overline{h'(z)}/h'(z) = \mu(z)$. This means that μ is a measurable form of the type $(-1, 1)$ with coefficient in the tangent bundle over S.

Let f be a quasiconformal homeomorphism of the surface S onto the surface S', and let $\tilde S$ and $\tilde S'$ be the coverings of these surfaces constructed from isomorphic subgroups of $\pi_1(S)$ and $\pi_1(S')$; then f lifts to a homeomorphism $\tilde f$ of the coverings if the diagram

$$\begin{array}{ccc} \tilde S & \xrightarrow{\tilde f} & \tilde S' \\ \pi \downarrow & & \downarrow \pi' \\ S & \xrightarrow{f} & S' \end{array}$$

is required to commute, where π and π' are the corresponding projections. Introducing on $\tilde S$ and $\tilde S'$ the natural complex structure in which the projections π and π' are holomorphic mappings, we get that the Beltrami differential

$\mu_f(z)\, d\bar{z}/dz$ on S lifts to a Beltrami differential $\mu_{\tilde{f}}(\varsigma)\, d\bar{\varsigma}/d\varsigma$ which satisfies the relation

$$\mu_{\tilde{f}}(\gamma(\varsigma))\overline{\gamma'(\varsigma)}/\gamma'(\varsigma) = \mu_{\tilde{f}}(\varsigma) \tag{22}$$

for all covering automorphisms γ of the surface $\tilde{S}$ (see §2 in Chapter II).

Quasiconformal mappings preserve the measurable structure on a Riemann surface, because the structure constructed by means of the metric (12) is equivalent to the original structure; but they do not preserve the conformal structure. For example, what is more, any two closed Riemann surfaces of the same genus and with the same number of deleted points are quasiconformally equivalent. However, not every two homeomorphic Riemann surfaces are quasiconformally equivalent: for instance, this is not so for the plane $\mathbf{C}$ and the disk U. In other words, quasiconformal equivalence is weaker than conformal equivalence, but is stronger than topological equivalence. In particular, it will be seen that quasiconformal mappings preserve the branching type over a given Riemann surface.

We also need to consider quasiconformal automorphisms f of $\overline{\mathbf{C}}$ that are compatible with general Kleinian groups G, i.e., such that $G_f = fGf^{-1}$ are again Kleinian groups. For this, it is again necessary that the form $\mu(z)\, d\bar{z}/dz$ be G-invariant, where $\mu = \mu_f$, and consequently, as in the case of (22),

$$\mu(\gamma(z))\overline{\gamma'(z)}/\gamma'(z) = \mu(z) \text{ for all } \gamma \in G \ (z \in \mathbf{C}). \tag{23}$$

Here it is also required that μ be defined on the limit set $\Lambda(G)$ of G if that set has positive measure. It is usual to set $\mu(z) = 0$ there. Sullivan [228] recently showed that this equality must be satisfied automatically for finitely generated groups of the first kind. Combining his results with results of Krushkal′ [135], we can get that this is so also for groups satisfying condition (63) (Chapter III, §2). But in the general case it is not so (see Examples 39 and 41).

A direct computation shows that under condition (23) the mapping $\gamma_f = f \circ \gamma \circ f^{-1}$ is a Möbius transformation if γ is. Consequently, the homeomorphism f induces an isomorphism $\chi_f \colon G \to G_f$ by the formula $\chi_f(\gamma) = \gamma_f$. Such isomorphisms χ_f will be called *quasiconformal deformations of the group G*. However, we sometimes use the same term for the mapping f itself, as well as for the image group G_f.

The theory of quasiconformal mappings provides one of the main ways of investigating Riemann surfaces and Kleinian groups along with applications of them.

3. ***Multi-dimensional quasiconformal mappings.*** In what follows we shall have to consider also mappings of multi-dimensional manifolds. In the general case, quasiconformality can be defined, for example, geometrically as follows.

Let M and M' be two n-dimensional Riemann manifolds with respective metrics ρ and ρ', and let f be a homeomorphism $M \to M'$. For each $p_0 \in M$

let

$$K(f;p_0) = \varlimsup_{r\to 0} \frac{\sup_{\rho(p,p_0)=r} \rho'(f(p), f(p_0))}{\inf_{\rho(p,p_0)=r} \rho'(f(p), f(p_0))}. \tag{24}$$

The homeomorphism f is said to be *quasiconformal* if $\sup_M K(f;p_0) = K(f) < \infty$ (cf. (16)).

In the case when M and M' are regions in $\overline{\mathbf{R}}^n$ this definition is equivalent to the condition that f be a homeomorphism in the Sobolev class $W_n^1(M)$ and that there exist a constant $K < \infty$ such that

$$|\operatorname{grad} f(x)|^n \leq K\|f'(x)\|, \tag{25}$$

where $\|f'(x)\|$ is the Jacobian of f. Many differentiability properties of planar mappings carry over to multi-dimensional quasiconformal mappings. However, the class of spatial quasiconformal mappings is fairly narrow, because the system of differential equations obtained for them, which is an analogue of (10), is strongly overdetermined. In particular, here there is no existence theorem as in the planar case, variational methods have so far not been developed, and, furthermore, there are no general conditions for a quasiconformal mapping to be compatible with a Kleinian group $G \subset \mathcal{M}_n$, i.e., conditions of the type (23). At present there is only the geometric variational method of Apanasov and Tetenov (see [36]–[38]) for constructing quasiconformal deformations of spatial Kleinian groups.

§5. Finitely generated Kleinian groups. Cohomology methods. Stability

1. ***Finitely generated Kleinian groups.*** In studying Kleinian groups it is necessary to impose various restrictions on them. On the whole, the most advanced results have been obtained for finitely generated groups on the plane. Some have already been presented in the preceding sections.

One of the main results in the theory of Kleinian groups is the ***Ahlfors finiteness theorem***, which asserts that *a finitely generated Kleinian group on the plane is a group of finite type* (the converse is not true). This result admits a quantitative refinement, which we now present.

Each component Δ of the set $\Omega(G)$ allows the introduction in it of a non-Euclidean Poincaré metric $ds^2 = \lambda_\Delta^2(z)|dz|^2$ which is induced by the universal holomorphic covering $h\colon U \to \Delta$ and is the only complete conformal Riemann metric of constant curvature -1. By the ***Gauss-Bonnet formula***, the non-Euclidean area of the surface Δ/G_Δ is

$$\iint\limits_{\Delta/G_\Delta} \lambda_\Delta^2(z)\,dx\,dy = 2\pi\left(2p - 2 + \sum_{j=1}^{n}(1 - 1/\nu_j)\right), \tag{26}$$

where $(p, n; \nu_1, \dots, \nu_n)$ is the signature of the surface Δ/G_Δ. It is this area that is considered below. It is clear from (26) that this area is completely determined by the topological type of the surface.

Bers [49] obtained the following ***area theorem***:

1. *If G has N generators and is not elementary, then*

$$\text{area}\,(\Omega(G)/G) \leq 4\pi(N-1), \tag{27}$$

where the inequality is sharp and equality is attained, for example, for Schottky groups.

2. *If G is a finitely generated function group with invariant component Ω_0, then*

$$\text{area}\,(\Omega(G)/G) \leq 2 \cdot \text{area}\,(\Omega_0/G). \tag{28}$$

Considering that (26) gives the lower estimate

$$\text{area}\,(U/\Gamma) \geq \pi/21, \tag{29}$$

for all (finitely generated) Fuchsian groups, we obtain from (27) an estimate for the number m of components of $\Omega(G)/G$ (i.e., the number of nonequivalent components of $\Omega(G)$): $m \leq 84(N-1)$). This estimate can be improved (see [18] and [59]), namely, $m \leq 18(N-1)$; but even this inequality admits various refinements.

The following result is due to Abikoff and Maskit [8].

THEOREM. *Each finitely generated Kleinian group can be formed from elementary groups, degenerate groups, and web groups by means of Maskit-Klein combination.*

Maskit had established previously that *every finitely generated function group can be obtained by a combination of elementary groups, quasi-Fuchsian groups, and degenerate groups.*

These deep results permit one in many cases to reduce the study of arbitrary finitely generated groups to simpler groups (assuming, of course, that the relevant properties are preserved under combination).

In conclusion, we briefly describe the non-quasi-Fuchsian finitely generated function groups with simply connected components, i.e., with a single invariant component. The main results here are due to Bers [51], Maskit [180], and Marden [168]. Such groups have come to be called *b-groups* (in connection with the fact that they were obtained as boundary points of the Teichmüller spaces of finitely generated Fuchsian groups of the first kind; therefore, *boundary groups*, which have the indicated property by definition, are formally distinguished from the b-groups). The b-groups turn out to be of three types: a) nondegenerate (or regular); b) partially degenerate; c) degenerate.

It has been proved that (in a definite sense) most boundary groups are degenerate, and for each Riemann surface of finite type there exists a continuum of nonconjugate (in $\mathcal{M}_2$) degenerate groups G such that $\Omega(G)/G = S$ [51].

The *nondegenerate b-groups* are defined to be those for which equality holds in (28). These groups admit an almost exhaustive geometric description and have by now been thoroughly studied. Each nondegenerate group G represents a main surface S_0 corresponding to the invariant component $\Omega_0 \subset \Omega(G)$, i.e., $S_0 = \Omega_0/G$, along with finitely many surfaces $S_1, \ldots, S_l$ (of finite type) obtained from S_0 by shrinking certain nonhomotopic loops on S_0 to points. Examples of such groups are given in Figures 2 and 65.

For nondegenerate groups we have the interesting ***uniqueness theorem*** first proved by Il′yashenko [105], [106]: *each such group is uniquely (to within conjugacy in $\mathcal{M}_2$) determined by the complementary surfaces $S_1, \ldots, S_l$, i.e., by the components of the quotient $(\Omega(G)\backslash\Omega_0)/G$.*(6)

Partially degenerate b-groups occupy an intermediate position between nondegenerate and degenerate groups and are Kleinian groups for which inequality holds in (28). They are obtained by Maskit combination of quasi-Fuchsian and degenerate groups. Therefore, many of the "pathologies" of degenerate groups are inherent to them.

2. ***Some remarks about automorphic forms and cohomology methods.*** We mention briefly other analytic methods in the theory of Kleinian groups and uniformization. An approach suggested by Poincaré and involving the consideration of *automorphic forms* has turned out to be very fruitful and important here.

Let Γ be a Kleinian group and let $q \geq 0$ be an integer. A meromorphic function $f(z)$ in $\Omega(\Gamma)$ is called a Γ-*automorphic form of weight* $-2q$ if, for all $\gamma \in \Gamma$,

$$f(\gamma(z)) = f(z)(\gamma'(z))^{-q}, \tag{30}$$

or, if γ has the form (1),

$$f((az+b)/(cz+d)) = f(z)(cz+d)^{2q}. \tag{30'}$$

In particular, for $q = 0$ we have the *automorphic functions.*

Such forms are obtained as follows: If Γ is not elementary and $\infty \in \Omega(\Gamma)$, then the series $\sum_{\gamma\in\Gamma} |\gamma'(z)|^q$ converges (at points $z \in \Omega(\Gamma)$, and uniformly on the corresponding compact sets) for integers $q \geq 2$, and along with it the Poincaré theta series

$$\Theta_q f(z) = \sum_{\gamma\in\Gamma} f(\gamma(z))\gamma'(z)^q$$

converge, where the f are meromorphic functions in $\Omega(\Gamma)$. The properties of the theta series have been investigated most thoroughly for Fuchsian groups on the plane.

To the automorphic forms with respect to Γ on the surfaces in $\Omega(\Gamma)/\Gamma$ there correspond meromorphic differentials of appropriate orders. Therefore,

(6) *Translation editor's note*: This is not correct as stated. Correct statements can be found in [6], [105] or [188] (see [6] or [188] for proof).

for finitely generated Kleinian groups the dimension of spaces of such forms can be computed with the aid of the classical Riemann-Roch theorem.

We dwell further on automorphic mappings of regions in space. Apanasov [32] constructed and studied theta series and automorphic forms (30) for Kleinian groups in space, and then with their help he studied mappings automorphic with respect to arbitrary Fuchsian groups in $\mathbf{R}^n$. It is not yet established whether these mappings are of *bounded distorsion*, i.e., whether they satisfy inequality (25) (see [210]). For Fuchsian groups with a fundamental polyhedron of finite hyperbolic volume, automorphic mappings with bounded distorsion (the analogue of analytic automorphic functions on the plane) were constructed by Martio and Srebro in [174].

Cohomology methods also play a large role in the contemporary theory of Kleinian groups. Let q be a positive integer, and let Π_{2q-2} be the vector space of polynomials of degree at most $2q-2$. Every Kleinian group Γ (and even an arbitrary subgroup of $\mathcal{M}_2$) acts from the right on Π_{2q-2} by the formula $p\gamma = p(\gamma(z))\gamma'(z)^{1-q}$, where $p \in \Pi_{2q-2}$ and $\gamma \in \Gamma$, and the equality $p(\gamma_1 \circ \gamma_2) = (p\gamma_1)\gamma_2$ holds.

A mapping $\sigma\colon \Gamma \to \Pi_{2q-2}$ is called a *cocycle* if

$$\sigma(\gamma_1 \circ \gamma_2) = \sigma(\gamma_1)\gamma_2 + \sigma(\gamma_2);$$

the cocycle $\gamma \to p_0\gamma - p_0$ serves as the coboundary of an element $p_0 \in \Pi_{2q-2}$. The quotient of the space (of equivalence classes) of cocycles by the space of coboundaries is the first *Eichler cohomology* group $H^1(\Gamma, \Pi_{2q-2})$. This group actually depends only on the conjugacy class of the group Γ in $\mathcal{M}_2$.

It is not hard to show that if Γ is a finitely generated group with N generators, then, for $q \geq 2$,

$$\dim H^1(\Gamma, \Pi_{2q-2}) \leq (2q-1)(N-1) \tag{31}$$

(while $\dim H^1(\Gamma, \mathbf{C}) \leq N$), with equality only for free groups.

The structure of finitely generated Kleinian groups from the point of view of their cohomology groups has been investigated by Ahlfors [20], Kra [121], and others. It is by cohomology methods that the finiteness and area theorems presented above are obtained.

3. *Stability of Kleinian groups.* Let Γ be a finitely generated Kleinian group with generators $\gamma_1 \dots, \gamma_N$. Consider the homomorphisms $\chi\colon \Gamma \to \mathcal{M}_2$ preserving the square of the trace of parabolic elements $\gamma \in \Gamma$, i.e., $\operatorname{tr}^2 \chi(\gamma) = \operatorname{tr}^2 \gamma = 4$. Such homomorphisms are said to be *admissible* and form a finite-dimensional affine algebraic variety $\operatorname{Hom}_\alpha(\Gamma, \mathcal{M}_2)$. Clearly, every $\chi \in \operatorname{Hom}_\alpha(\Gamma, \mathcal{M}_2)$ is completely determined by the elements $\chi(\gamma_1), \dots, \chi(\gamma_N)$. Say that χ is *close to the identity* if the element $\chi(\gamma_j)$ is close to the element γ_j itself for $j = 1, \dots, N$ (for example, in the corresponding matrix representation). The group Γ is said to be (*quasiconformally*) *stable* if every admissible homomorphism of it sufficiently close to the identity is induced by a quasiconformal automorphic $f: \overline{\mathbf{C}} \to \overline{\mathbf{C}}$ with small $k(f) = \|\mu_f\|_\infty$. But if only the homomorphisms χ_f that are quasiconformal deformations of Γ have the indicated property, then Γ is said to be *conditionally stable.*

Stability of Γ means that some neighborhood of the identity homomorphism (i.e., of the group Γ) in $\mathrm{Hom}_\alpha(\Gamma, \mathcal{M}_2)$ is filled by quasiconformal deformations of Γ. The concept of stability, which was introduced by Bers, turns out to be very essential for the study of spaces of deformations of Kleinian groups, in particular, their complex analytic structure.

Questions of stability of Kleinian groups have been studied in [5], [51], [80], [130], [168], and elsewhere. In particular, Marden established that *all torsion-free geometrically finite Kleinian groups are stable.* For example, quasi-Fuchsian groups, Schottky groups, nondegenerate *b*-groups, and others are stable. Nonstable groups do exist (see Example 47). Krushkal′ [130], [136] proved that *all finitely generated Kleinian groups are conditionally stable.*

Sullivan [301] recently showed that *a Kleinian group* Γ *is stable if and only if it is geometrically finite*; this gives a necessary and sufficient condition for stability.

CHAPTER II

Uniformization of Riemann Surfaces and Manifolds

§1. What is uniformization?

This section serves as an introduction for the whole chapter and expounds the program according to which one of the most important directions of analysis and topology has evolved and continues to evolve.

The concept of uniformization, which is one of the main concepts in complex analysis and other areas of mathematics, was introduced in the classical literature as far back as the second half of the last century when a detailed study of multi-valued analytic functions was beginning. In its original understanding, to *uniformize* a given multi-valued analytic function $w = F(z)$ (or the corresponding analytic expression or set) means to represent this function parametrically with the help of single-valued (uniform) holomorphic or, in general, meromorphic functions $z = z(t)$, $w = w(t)$ such that $w(t) = F[z(t)]$.

A more precise definition is the following. A set $A \subset \mathbf{C}^m$ (or $\mathbf{CP}^m$) is said to be *uniformized by a system* $f = (f_1, \ldots, f_m)$ *of functions meromorphic in a region* $D \subset \mathbf{C}^n$ if f is a holomorphic covering $D_0 \to A_0$ of a dense subset A_0 of A ($D_0 \subset D$) with discrete fiber on which a (discrete) group G of automorphisms of D acts transitively.

For the present we discuss mainly uniformization of complex algebraic or more general analytic curves, i.e., Riemann surfaces (one-dimensional complex manifolds with a conformal structure). In §§7 and 8 we give some results obtained for multi-dimensional real manifolds admitting a conformal structure.

Let us consider some elementary examples.

EXAMPLE 6. The circle $x^2 + y^2 = 1$ (and the corresponding complex curve in $\mathbf{C}^2$) is uniformized by the trigonometric functions $x = \cos t$, $y = \sin t$ or by the *rational* functions $x = (1 - t^2)/(1 + t^2)$, $y = 2t/(1 + t^2)$. In the first case $D_0 = \mathbf{C}$ and G is the group of translations $t \to t + 2k\pi$, $k \in \mathbf{Z}$, and in the second case $D_0 = \mathbf{C} \setminus \{i, -i\}$ and G is the trivial group.

EXAMPLE 7. The cubic curve

$$w^2 = a_0z^3 + a_1z^2 + a_2z + a_3 \tag{32}$$

does not admit a rational parametrization, but it can be uniformized with the help of *elliptic* functions. Namely, by a linear change in w and z the curve (32) can be brought to the form

$$w^2 = 4(z - e_1)(z - e_2)(z - e_3), \tag{33}$$

where e_1, e_2 and e_3 are distinct numbers with $e_1 + e_2 + e_3 = 0$; then, taking the Weierstrass $\wp$-function

$$\wp(\varsigma) = \frac{1}{\varsigma^2} + \sum_{m,n}\left[\frac{1}{(\varsigma - m\omega_1 - n\omega_2)^2} - \frac{1}{(m\omega_1 + n\omega_2)^2}\right]$$

with corresponding periods ω_1 and ω_2 (so that $\operatorname{Im}(\omega_2/\omega_1) > 0$), where $(m, n) \in \mathbf{Z}^2\backslash\{0, 0\}$, we can write $z = \wp(\varsigma)$ and $w = \wp'(\varsigma)$ by virtue of (33); in this case G is a free abelian group of rank 2.

Beginning with Riemann, many of the leading mathematicians in the second half of the last century occupied themselves with the uniformization problem for an arbitrary algebraic curve defined by a general algebraic equation

$$P(z, w) = \sum_{j,k} a_{jk}z^jw^k = 0, \tag{34}$$

where P is an irreducible polynomial over the field $\mathbf{C}$. The problem is again to find a parametric representation of all the pairs (z, w) satisfying equation (34) by means of single-valued analytic functions of a complex variable. This problem arose even earlier due, for example, to the needs of analysis in connection with integration of the corresponding algebraic functions $w(z)$ determined by such equations. More generally, a rational function $R(z, w)$ of the variables z, w (where $w = w(z)$) is connected with equation (34), and the problem arises of studying the so-called abelian integrals $\int R(z, w)\,dz$; to solve this it is again necessary to have a parametric representation $z = z(t)$, $w = w(t)$.

Associated with each algebraic curve in a definite way (for example, the way indicated below) there is a certain nonnegative integer p called the *genus* of this curve. It turns out that *all curves of genus* 0 *can be uniformized by rational functions, all those of genus* 1 *can be uniformized by elliptic functions, and all those of genus* $p > 1$ *can be uniformized by meromorphic functions defined on proper open subsets of* $\mathbf{C}$, *for example, in the disk* (in other words, *for* $p > 1$ *the set of solutions of any polynomial equation* (34) *can be uniformized by appropriate meromorphic functions in the disk*).

This is one of the deepest results in mathematics as a whole, and has far-reaching generalizations. It is due to Klein, Poincaré, and Koebe.

If in the set of pairs (z, w) in $\mathbf{C}^2$ satisfying (34) we introduce an analytic structure with the help of elements of the corresponding algebraic function $w(z)$ (or $z(w)$), then we get the Riemann surface of this function. The surface is compact and is a finite-sheeted covering sphere. It turns out that, *up to a conformal equivalence* (a homeomorphism preserving complex structure), *all compact Riemann surfaces are obtained in this way*; moreover, the coordinates of the points on the curve are meromorphic functions on the Riemann surface. Consequently, it is necessary to know how to uniformize Riemann surfaces.

Poincaré posed the problem of uniformizing the set of solutions of arbitrary analytic equations (in other words, uniformizing general analytic curves and even manifolds) when the P in (34) is not a polynomial but some convergent power series in two variables considered in its whole domain of existence, i.e., together with all possible analytic continuations of it. The problem of uniformizing algebraic and arbitrary analytic manifolds was included by Hilbert among the problems posed in his famous report at the Second International Congress of Mathematicians in 1900 in Paris.

A complete solution to the uniformization problem has not yet been obtained (with the exception of the one-dimensional case); however, there have been essential advances here with have served as foundations for the development of many important ideas and methods in mathematics. This applies, in particular, to topological methods, covering spaces, existence theorems for partial differential equations, existence and distorsion theorems for conformal mappings, and so on.

The theory of quasiconformal mappings has served as an important (and in many cases basic) tool for uniformization of Riemann surfaces. It has led to results like the simultaneous uniformization of several and even of *all* Riemann surfaces of a given genus. This question is connected with Teichmüller spaces and the general theory of Kleinian groups.

The elementary examples considered above already reveal *the essence of uniformization: it is necessary to construct a universal covering for a given Riemann surface on which* (*since it is simply connected*) *the corresponding functions are single-valued.*

Construction of the covering leads to the consideration of planar regions and corresponding discrete groups of conformal automorphisms of these regions. It turns out that for a suitable choice of regions these automorphisms are *Möbius* mappings, and hence the groups are *Kleinian* and even have an invariant simply connected component. If Δ denotes this component and G the group itself, then the original (uniformizable) Riemann surface S is conformally equivalent to the surface Δ/G. Any multi-valued analytic function f on S becomes single-valued on S in terms of the universal holomorphic covering $\pi\colon \Delta \to S$, i.e., the function $f \circ \pi$. Therefore, a uniformization of the Riemann surface is often understood to be the pair (G, Δ), or even the Kleinian group G itself, and it is said that G *uniformizes* S. We shall understand uniformization in just this sense.

§2. The fundamental group and covering spaces. Planar coverings

As already mentioned, the theory of uniformization and the theory of Kleinian groups are permeated with the concepts of a covering space and the fundamental group. Therefore, we recall these most important constructions in this section.

1. ***The fundamental group.*** Let X be a topological space. A *path* in X is defined to be a continuous mapping f of the closed interval $I = [0, 1]$ into X. A path whose endpoints $f(0)$ and $f(1)$ coincide is called a *closed path* or *loop.* If $f(0) = f(1) = x_0 \in X$, then we speak of a *loop* at the point x_0.

Two paths $f_1, f_2: I \to X$ are said to be *homotopic* if there exists a continuous mapping $F: I \times I \to X$ such that $F(t, 0) = f_1(t)$ and $F(t, 1) = f_2(t)$ for all $t \in I$, and $F(0, s) = f_1(0) = f_2(0)$ and $F(1, s) = f_1(1) = f_2(1)$ for all $s \in I$. It is not hard to verify that the relation of being homotopic is an equivalence relation. We denote it by the symbol $\simeq$, and the equivalence class of a path f by $[f]$.

The product of two loops $f_1, f_2: I \to X$ at the same point is defined by

$$(f_1 \cdot f_2)(t) = \begin{cases} f_1(2t), & 0 \le t \le 1/2, \\ f_2(2t - 1), & 1/2 \le t \le 1. \end{cases}$$

Observe that if $f_1 \simeq f_2$ and $g_1 \simeq g_2$, then $f_1 \cdot g_1 \simeq f_2 \cdot g_2$.

Suppose now that x_0 is an arbitrary particular point in X, and consider the set of homotopy classes of loops at x_0. It is easy to check that this set is a group if the multiplication operation on it is given by the formula $[f_1] \cdot [f_2] = [f_1 \cdot f_2]$. This group is called the *fundamental group of* X *at* x_0 and denoted by $\pi_1(X, x_0)$. As is clear from the definition, the fundamental group of X depends on the choice of the point x_0. It is not hard to show that if X is arcwise connected, then for any two points $x_0, y_0 \in X$ the groups $\pi_1(X, x_0)$ and $\pi_1(X, y_0)$ are isomorphic. Therefore, in the case of arcwise connected spaces we can speak of the abstract *fundamental group* $\pi_1(X)$ *of* X.

It can be verified directly that every continuous mapping $\varphi: X \to Y$ induces a homomorphism $\varphi_*: \pi_1(X, x_0) \to \pi_1(Y, \varphi(x_0))$ by the rule $\varphi_*([f]) = [\varphi \circ f]$. In particular, if φ is a homeomorphism, then the groups $\pi_1(X)$ and $\pi_1(Y)$ are isomorphic.

2. ***Coverings.*** Again, let X be a topological space. A *covering space* over X, or a *covering*, is defined to be a pair $(\tilde{X}, \pi)$ consisting of a space $\tilde{X}$ and a continuous mapping $\pi: \tilde{X} \to X$ satisfying the condition that each $x \in X$ has an arcwise connected (open) neighborhood U such that any connected component of the set $\pi^{-1}(U)$ is mapped homeomorphically onto U under the mapping π. The mapping π is called a *covering mapping* or *projection.*

Covering spaces are closely connected with fundamental groups. The following assertions establish a connection between the fundamental group of a space X and the fundamental group of a space covering it.

1) *For any path* $f: I \to X$ *with initial point* x_0 *there exists a unique path* $\tilde{f}: I \to \tilde{X}$ *with initial point* $\tilde{x}_0 \in \pi^{-1}(x_0)$ *such that* $\pi \circ \tilde{f} = f$ *(lifting of paths).*

2) *If* g_1 *and* g_2 *are paths starting at the same point in the covering space* $\tilde{X}$ *and such that* $\pi \circ g_1 \simeq \pi \circ g_2$, *then also* $g_1 \simeq g_2$ *(the covering homotopy property).*

3) *The homomorphism* $\pi_*: \pi_1(\tilde{X}, \tilde{x}_0) \to \pi_1(X, x_0)$ *induced by the projection* $\pi: \tilde{X} \to X$, *where* $x_0 = \pi(\tilde{x}_0)$, *is a monomorphism.*

4) *Let* $\tilde{x}_0$ *and* $\tilde{x}_1$ *be points in* $\tilde{X}$ *with the same projection* x_0. *Then the images of the groups* $\pi_1(\tilde{X}, \tilde{x}_0)$ *and* $\pi_1(\tilde{X}, \tilde{x}_1)$ *under the mapping* π_* *are conjugate subgroups of* $\pi_1(X, x_0)$; on the other hand, each subgroup in the class of subgroups conjugate to $\pi_*(\pi_1(\tilde{X}, \tilde{x}_0))$ is obtained as the image $\pi_*(\pi_1(\tilde{X}, \tilde{x}_1))$ for some point $\tilde{x}_1 \in \pi^{-1}(x_0)$. Consequently, *the images* $\pi_*(\pi_1(\tilde{X}, \tilde{x}))$ *over all* $\tilde{x} \in \pi^{-1}(x_0)$ *form a complete class of conjugate subgroups in the group* $\pi_1(X, x_0)$.

Two coverings $(\tilde{X}', \pi')$ and $(\tilde{X}'', \pi'')$ over X are said to be *equivalent* if there exists a homeomorphism $\varphi: \tilde{X}' \to \tilde{X}''$ such that the diagram

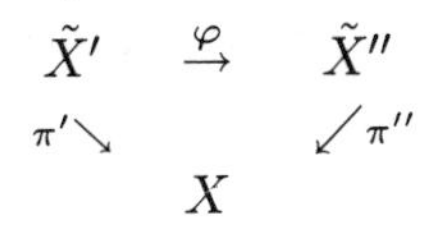

commutes.

5) *Two covering spaces* $(\tilde{X}', \pi')$ *and* $(\tilde{X}'', \pi'')$ *are equivalent if and only if, for any two points* $\tilde{x}' \in \tilde{X}'$ *and* $\tilde{x}'' \in \tilde{X}''$ *such that* $\pi'(\tilde{x}') = \pi''(\tilde{x}'') = x$, *the subgroups* $\pi'_*(\pi_1(\tilde{X}', \tilde{x}'))$ *and* $\pi''_*(\pi_1(\tilde{X}'', \tilde{x}''))$ *belong to the same class of conjugate subgroups of* $\pi_1(X, x)$.

A homeomorphism $f: \tilde{X} \to \tilde{X}$ is called a *covering homeomorphism* or a *covering transformation* if the diagram

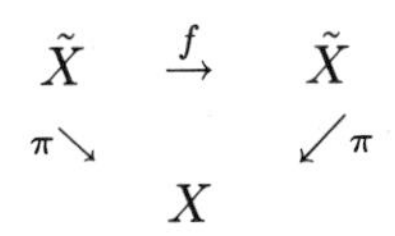

commutes.

The set of covering homeomorphisms of a covering space (X, π) forms a group under superposition of mappings; denote it by $G(\tilde{X}, X)$.

6) *For any points* $x \in X$ *and* $\tilde{x} \in \pi^{-1}(x)$ *the group* $G(\tilde{X}, X)$ *is isomorphic to the factor group* $N[\pi_*(\pi_1(\tilde{X}, \tilde{x}))]/\pi_*(\pi_1(\tilde{X}, \tilde{x}))$, *where* $N[\pi_*(\pi_1(\tilde{X}, \tilde{x}))]$ *denotes the normalizer of the subgroup* $\pi_*(\pi_1(\tilde{X}, \tilde{x}))$ *in* $\pi_1(X, x)$.

Especially important is the case of covering spaces such that $\pi_*(\pi_1(\tilde{X}, \tilde{x}))$ is a normal subgroup of $\pi_1(X, x)$. Such covering spaces are called *regular*. It can be shown that a covering $(\tilde{X}, \pi)$ is regular if and only if all liftings to $\tilde{X}$ of any loop at an $x \in X$ with initial point at an arbitrary point $\tilde{x} \in \pi^{-1}(x)$ are either closed or open paths.

7) *If $(\tilde{X}, x)$ is a regular covering over X, then for any points $x \in X$ and $\tilde{x} \in \pi^{-1}(x)$ the group $G(\tilde{X}, X)$ is isomorphic to $\pi_1(X, x)/\pi_*(\pi_1(\tilde{X}, \tilde{x}))$.*

A covering $(\tilde{X}, \pi)$ of a space X is said to be *universal* if $\tilde{X}$ is simply connected, i.e., $\tilde{X}$ is connected and $\pi_1(\tilde{X})$ is the trivial group.

8) *If $(\tilde{X}, \pi)$ is a universal covering over X, then the group $G(\tilde{X}, X)$ is isomorphic to $\pi_1(X)$, and the order of $\pi_1(X)$ is equal to the number of sheets of the covering $(\tilde{X}, \pi)$* (the latter number is defined to be the common cardinality of the fibers $\pi^{-1}(x)$, which does not depend on the choice of the point $x \in X$).

9) *The group $G(\tilde{X}, X)$ acts on the fibers $\pi^{-1}(x)$, $x \in X$, transitively if and only if $(\tilde{X}, \pi)$ is a regular covering.*

Consequently, if the covering $(\tilde{X}, \pi)$ of X is regular, then there is a natural homeomorphism of X onto $\tilde{X}/G(\tilde{X}, X)$.

10) *If $(\tilde{X}, \pi)$ is a regular covering over X, then $G(\tilde{X}, X)$ acts on $\tilde{X}$ freely* (without fixed points), *and also discontinuously*, i.e., each point $\tilde{x} \in \tilde{X}$ has a neighborhood U such that $\varphi(U) \cap U = \varnothing$ for all $\varphi \in G(\tilde{X}, X)\backslash\{\mathbf{1}\}$.

If X is arcwise connected and G is a discontinuous group of homeomorphisms of it that acts freely, then (X, π) is a regular covering over X/G, and $G = G(X, X/G)$.

The following assertion is one of the main results in the theory of covering spaces.

11) *Let X be a topological manifold (i.e., a locally Euclidean Hausdorff space with a countable base). Then for any class $\mathfrak{A}$ of conjugate subgroups of $\pi_1(X, x)$ there exists a covering space $(\tilde{X}, \pi)$ corresponding to a given class $\mathfrak{A}$, i.e., such that $\pi_*(\pi_1(\tilde{X}, \tilde{x})) \in \mathfrak{A}$.*

This implies that if H is an arbitrary subgroup of $\pi_1(X, x_0)$, then there exists a covering $(\tilde{X}, \pi)$ over X such that $\pi_*(\pi_1(\tilde{X}, \tilde{x}_0)) = H$ for some point $\tilde{x}_0 \in \pi^{-1}(x_0)$.

The subgroup H is called a *defining subgroup of the covering* $(\tilde{X}, \pi)$, and the covering $(\tilde{X}, \pi)$ itself is said to be *constructed from the subgroup* H.

3. ***Regular planar coverings.*** In this section a *surface* is understood to be a two-dimensional topological manifold without boundary. A surface is said to be *planar* if it is homeomorphic to some region on the two-dimensional sphere. A covering $\pi: \tilde{S} \to S$ of a surface S is called a *planar covering* if S is a planar surface.

We shall now see that regular planar coverings are connected in a natural way with uniformization of Riemann surfaces by Kleinian groups.

Let G be a Kleinian group on the plane having an invariant component Δ, and let $S = \Delta/G$. Then, as noted in subsection 2, the natural projection

$\pi: \Delta \to S$ is a regular covering, in this case planar, and G is the group of covering homeomorphisms of this covering. Now let T be a Riemann surface and let $h: S \to T$ be a homeomorphism; then $h \circ \pi: \Delta \to T$ is obviously a regular covering of T. The pair (G, Δ) is called a *topological uniformization* of the surface T. If h is a conformal homeomorphism, then (G, Δ) is called a *uniformization* of T.

Each Riemann surface T admits such a uniformization. Namely, let $\pi: \tilde{T} \to T$ be an arbitrary regular planar covering of the Riemann surface T. Then $\tilde{T}$ is endowed in the natural way with a conformal structure in which the projection π is holomorphic, and the group $G(\tilde{T}, T)$ of covering homeomorphisms of this covering becomes the group of conformal automorphisms of $\tilde{T}$.

We have the ***theorem of Maskit*** [178], which asserts that *there exists a conformal homeomorphism $f: \tilde{T} \to \Delta$, where Δ is a region on the Riemann sphere, such that $fG(\tilde{T}, T)f^{-1} = G$ is a Kleinian group with invariant component Δ*. Thus, *every Riemann surface T can be uniformized by a Kleinian group*. Consequently, *the problem of describing all the uniformizations of a given Riemann surface S reduces to the problem of describing all the regular planar coverings over S*.

It turns out that each regular planar covering can be characterized by some set of simple pairwise disjoint loops on the surface S.

Let $v_1, \dots, v_n, \dots$ be a set of loops on S at points $o_1, \dots, o_n, \dots$, let u_i be some path joining o_i to a particular point o, and let f_i be the element of $\pi_1(S, o)$ containing the loop $u \cdot v_i \cdot u_i^{-1}$(7). Denote by N the smallest normal subgroup of $\pi_1(S, o)$ containing all the elements f_i, and consider the regular covering $\pi: \tilde{S} \to S$ of S whose defining subgroup coincides with N. It is easy to verify that the covering $(\tilde{S}, \pi)$ determined in this way depends only on the free homotopy classes of the loops $v_1, \dots, v_n, \dots$ (i.e., the classes of homotopic loops without specification of a particular initial point); the normal subgroup N is unambiguously defined for any point $o \in S$ and also depends only on the free homotopy classes of the loops $v_1, \dots, v_n, \dots$. In this case we say that *the normal subgroup N is spanned by the loops $v_1, \dots, v_n, \dots$*, and we write $N = \langle v_1, \dots, v_n, \dots \rangle$.

The regular planar coverings over a surface S are completely described by the following two ***theorems of Maskit*** [176]:

I. *Let S be an orientable surface, and $v_1, \dots, v_n, \dots$ a set of simple disjoint loops on S. If $(\tilde{S}, \pi)$ is the regular covering with defining subgroup $N = \langle v_1^{\alpha_1}, \dots, v_n^{\alpha_n}, \dots \rangle$, where $\alpha_1, \dots, \alpha_n, \dots$ are positive integers, then $\tilde{S}$ is a planar surface.*

II (***Theorem on planar coverings***).*Let $(\tilde{S}, \pi)$ be a regular planar covering of an orientable surface S with defining subgroup N. If S is a surface of finite type, i.e., $\pi_1(S)$ is finitely generated, then there exists finitely many*

(7) *Translation editor's note*: The correct expression is $u_i \cdot v_i \cdot u_i^{-1}$.

simple disjoint loops $v_1, \ldots, v_n$ *and positive integers* $\alpha_1, \ldots, \alpha_n$ *such that* $\langle v_1^{\alpha_1}, \ldots, v_n^{\alpha_n} \rangle = N$.

An application of these theorems of Maskit to uniformization of Riemann surfaces by Kleinian groups will be given in §4.

§3. Uniformization of Riemann surfaces

We now present the main results on uniformization of one-dimensional complex manifolds.

The ***Klein-Poincaré uniformization theorem***, proved in the general case by Poincaré, asserts that *each Riemann surface* S *can be represented (to within a conformal equivalence) in the form* $\hat{S}/\Gamma$, *where* $\hat{S}$ *is one of the three canonical regions, namely, the extended plane* $\overline{\mathbf{C}}$ *(the sphere), the plane* $\mathbf{C}$, *or the disk* U, *and* Γ *is a discrete group of Möbius automorphisms of* $\hat{S}$ *acting freely there (without fixed points) and determined to within conjugation in the Möbius group* $\mathcal{M}_{\hat{S}}$ *of automorphisms of* $\hat{S}$.

The cases $\hat{S} = \overline{\mathbf{C}}$, $\mathbf{C}$, and U are mutually exclusive. Surfaces S with such coverings are said to be of *elliptic, parabolic*, and *hyperbolic type*, respectively. It is not hard to show that $\hat{S} = \overline{\mathbf{C}}$ only in the case when S itself is conformally equivalent to the sphere (and, hence, Γ is trivial); $\hat{S} = \mathbf{C}$ when S is conformally equivalent to $\mathbf{C}$ or $\mathbf{C}\backslash\{0\}$ or a torus, and, correspondingly, Γ is trivial or a translation group with generator $z \to z + \omega$ ($\omega \in \mathbf{C}\backslash\{0\}$) or a free abelian group of rank 2 with generators $z \to z + \omega_1$ and $z \to z + \omega_2$, where $\omega_1, \omega_2 \neq 0$ are complex numbers with $\operatorname{Im}(\omega_2/\omega_1) > 0$. With the exception of these elementary cases, S is equivalent to the quotient of the disk by a torsion-free Fuchsian group.

The canonical projection $\pi: \hat{S} \to \hat{S}/\Gamma$ is an unbranched covering and uniformizes all the functions on S, since the functions $f \circ \pi$ are single-valued on $\hat{S}$.

The theorem given above admits generalization also to surfaces with branching. Let us associate with the points $p \in S$ a function $\nu_\Sigma(p)$ equal to zero outside some discrete set $\Sigma \subset S$ and taking positive integer values or the value ∞ on Σ; assume also that if $S = \overline{\mathbf{C}}$, then Σ must not reduce to a single point, while if Σ consists of two points p_1, p_2, then $\nu_\Sigma(p_1) = \nu_\Sigma(p_2)$.

In the general form the ***Klein-Poincaré uniformization theorem*** asserts that *every Riemann surface* S *with a discrete set* Σ *distinguished on it and a given function* $\nu_\Sigma(p)$ *can be represented up to conformal equivalence in the form* $S = \hat{S}/\Gamma$, *where* $\hat{S} = \overline{\mathbf{C}}$, $\mathbf{C}$, *or* U, Γ *is a discrete group of automorphisms of* S, *and the covering* $\hat{S} \to \hat{S}/\Gamma$ *is branched over* S *at the points* $p \in \Sigma$ *and has there orders* $\nu_\Sigma(p)$.

The fixed points of stationary elliptic subgroups of Γ of orders $\nu_\Sigma(p)$ correspond to the branch points p.

There are various proofs of this fundamental theorem. Their scheme is as follows: Using the topological method indicated in §2, one constructs a universal covering $\hat{S}_0$ of the surface $S_0 = S \setminus \{p \in \Sigma : \nu_\Sigma(p) = \infty\}$ such that the covering $\pi: \hat{S}_0 \to S_0$ has the required branching orders over the points $p \in \Sigma$ (when such points are present). Then the conformal structure is lifted from S_0 to $\hat{S}_0$ as a covering, so that the covering π becomes holomorphic and reduces (to within equivalence) to taking the quotient of $\hat{S}_0$ by a group Γ of conformal automorphisms of it. Since the surface $\hat{S}_0$ is simply connected, we can use the Riemann theorem on conformal mapping of simply connected Riemann surfaces onto canonical regions.

A somewhat different approach to the solution of the uniformization problem was proposed roughly at the same time by Koebe. The ***general uniformization principle***, going back to Koebe, asserts that *if a Riemann surface $\tilde{S}$ is topologically equivalent to a planar region D, then there also exists a conformal homeomorphism of $\tilde{S}$ onto D.* The same problem of (analytic) uniformization reduces to the topological problem of finding all the (branched, in general) planar coverings $\tilde{S} \to S$ of a given Riemann surface S. The solution of this topological problem was presented in §2 and is given by the theorem of Maskit.

In particular, if S is a closed orientable surface of genus $p \geq 1$, then its fundamental group is generated by $2p$ elements:

$$\pi_1(S) = \left\{ a_1, b_1, \ldots, a_p, b_p : \prod_{j=1}^{p} [a_j, b_j] = 1 \right\}$$

(where [,], as usual, is the commutator), and the normal subgroup generated by $a_1, \ldots, a_p$ can be taken to be the normal subgroup N defining $\tilde{S}$. Then *the surface S can be uniformized by a Schottky group G* of genus p—a free purely loxodromic Kleinian group with p generators: $S = \Omega(G)/G$. This is the classical ***theorem on cuts***, as proved by Koebe.

We shall discuss the subsequent development of Koebe's method in the next section. For the present we note that, as pointed out in §2, *uniformization of Riemann surfaces is actually exhausted by the Kleinian groups.*

With the help of quasiconformal mappings it is possible to obtain a uniformization theorem of more general character, namely, to prove the possibility of *simultaneously uniformizing* several surfaces. This was done by Bers (see [31]). For example, we present the following general theorem due to him:

Let G be a Kleinian group, and suppose that $\Omega(G)/G = \bigcup_j S_j$, where the S_j are Riemann surfaces with branching, and let $f_j: S_j \to S'_j$ be quasiconformal mappings with $K(f_j) \leq K < \infty$. Then there exists a quasiconformal deformation of the group G into a group $G' \subset \mathcal{M}_2$ such that $\Omega(G')/G' = \bigcup_j S'_j$.

This theorem can be proved by the following scheme: The Beltrami differentials μ_{f_j} are canonically lifted from S_j to $\Omega(G)$, and then are extended to a single Beltrami G-differential μ by the equality $\mu|_{\Lambda(G)} = 0$. The group G_{f^μ} has all the required properties.

The preceding uniformization theorems can be obtained from this by an appropriate choice of G. As another corollary we mention the following result: *If S and S' are two quasiconformally equivalent Riemann surfaces of hyperbolic type with degenerate boundary components (i.e., without ideal boundary curves), then there exists a quasi-Fuchsian group G of the first kind such that* $\Omega(G)/G = S \cup S'$. This result can be strengthened to obtain the uniqueness of G in $\mathcal{M}_2$ by singling out homotopy classes of quasiconformal homeomorphisms of S onto S' (cf. §5), and it can also be generalized to surfaces with nondegenerate boundaries.

In §5 we show how to simultaneously uniformize all surfaces in a collection of quasiconformally equivalent surfaces.

§4. Uniformization by Kleinian groups

In this section we give a construction and a description of *all* the uniformizations of Riemann surfaces by Kleinian groups. These results are due mainly to Koebe and Maskit.

Let S be a Riemann surface, (G, Δ) a uniformization of it, $\pi: \Delta \to \Delta/G$ the canonical projection, and $h: \Delta/G \to S$ an arbitrary particular conformal homeomorphism. Consider a point $x \in S$. It is not hard to verify that $\pi: \Delta \to \Delta/G$ is a regular branched covering (see §9), so the stabilizers G_{z_1} and G_{z_2} in G of any two points $z_1, z_2 \in (h \circ \pi)^{-1}(x)$ are finite groups of the same order. This allows us to define unambiguously the *branching order* $\nu = \nu(x)$ of the point x to be the order of the stabilizer G_z of an arbitrary point $z \in (h \circ \pi)^{-1}(x)$. A point $x \in S$ is called a *branch point* of S if $\nu(x) > 1$.

It is assumed everywhere in this section that all the Kleinian groups considered are finitely generated. Then, by the Ahlfors finiteness theorem, S is a finite Riemann surface with a finite number of branch points.

Consider a Riemann surface S of finite type (p, k), i.e., a closed Riemann surface $\overline{S}$ of genus p with k deleted points. The *genus* of S of genus p is defined to be the genus of $\overline{S}$.

Let (G, Δ) be a uniformization of S, let $\overline{S} \backslash S = \{y_1, \dots, y_k\}$, and let $x_1, \dots, x_m$ be branch points of S of respective orders $\nu_1, \dots, \nu_m$ ($2 \le \nu_j < \infty$, $j = 1, \dots, m$). We also say that (G, Δ) uniformizes the surface $\overline{S}$, and regard the points $y_1, \dots, y_m$ as branch points of $\overline{S}$ of infinite order. The vector $(g, n; \nu_1, \dots, \nu_n)$ is called the *signature of the uniformization* (G, Δ); here $n = k + m$, and some of the ν_j may be equal to ∞.

It follows from obvious topological considerations that vectors $(0, 1; \nu)$ and $(0, 2; \nu_1, \nu_2)$, $\nu_1 \neq \nu_2$, cannot be signatures of any uniformization. But the remaining vectors of this form are always signatures, on the strength of the Klein-Poincaré uniformization theorem.

The Kleinian group Γ constructed in this theorem is called the *branched universal covering group of the surface* S. This group is completely determined by the signature $(\Gamma, \hat{S})$ in the sense that it will be Fuchsian or elementary. To describe more complicated uniformizations it is necessary to consider a certain class of subgroups of Kleinian groups with an invariant component.

Let G be a Kleinian group with an invariant component Δ. If H is a subgroup of G, then $\Omega(H) \supset \Omega(G)$, and thus H has an invariant component $\Delta(H)$ which contains Δ; this component $\Delta(H)$ is called the *marked component* of H.

The subgroup H is called a *factor subgroup* of G if 1) $\Delta(H)$ is simply connected, 2) H does not contain accidental parabolic elements, 3) each parabolic element of G with fixed point in $\Lambda(H)$ belongs to H, and 4) H is a maximal subgroup of G satisfying 1)–3).

Each factor subgroup of a branched universal covering group G coincides with G itself. All the Schottky factor subgroups are trivial, and conversely. Maskit [188] showed that *all the factor subgroups of a Kleinian group G are finitely generated, and, moreover, they are exhausted by the elementary, the degenerate, and the quasi-Fuchsian groups.* There is a close connection between the factor subgroups of G and the topology of the planar (branched, in general) covering $\pi\colon \Delta \to \Delta/G = S$.

Let H be a subgroup of G. A set $A \subset \overline{\mathbf{C}}$ is said to be *precisely invariant under H in G* if $\gamma(A) = A$ for all $\gamma \in H$ and $\gamma(A) \cap A = \varnothing$ for all $\gamma \in G \backslash H$.

Let $Y \subset S$ be a connected set, A a connected component of $\pi^{-1}(Y)$, and H the stabilizer of A in G. Then A is precisely invariant under H in G. It will be said that *A covers Y*, and H is a *covering subgroup* for Y.

The next theorem shows the connection between the factor subgroups of G and the precisely invariant sets.

DECOMPOSITION THEOREM [184]. *Let G be a nonelementary Kleinian group with an invariant component Δ that uniformizes a Riemann surface S with signature $(p, n; \nu_1, \dots, \nu_n)$. Then there exists a set of simple disjoint loops $v_1, \dots, v_m$, $0 \le m \le 3p - 3 + n$, on $S \backslash \{x_1, \dots, x_n\}$ (where the x_i are branch points of order ν_i on S) that divide S into connected components $Y_1, \dots, Y_s$, $1 \le s \le 2p - 2 + n$, in such a way that the following conditions are satisfied:*

a) *each factor subgroup of G is a covering subgroup for some Y_i;*

b) *each covering subgroup for any Y_i is a factor subgroup of G;*

c) *two factor subgroups are conjugate in G if and only if they cover the same Y_i; and*

d) *each covering subgroup for any loop v_j is the trivial group, or a cyclic elliptic group, or a cyclic parabolic group.*

We note two special cases of this theorem. If $m = 0$, then G is an elementary group with a simply connected set of discontinuity, or a degenerate group, or a quasi-Fuchsian group. Another limit case is that of a Schottky group.

In this case, $n = 0$ and $m = p$, the loops $v_1, \ldots, v_p$ are homologically independent (as one-dimensional cycles), there exists only one surface Y, which is homeomorphic to the sphere with $2p$ disks removed, each covering subgroup for Y is trivial, and each component A of the set $\pi^{-1}(Y)$ is a fundamental region of G.

A uniformization (G, Δ) of a Riemann surface S is said to be *standard* if the group G is torsion-free and does not contain accidental parabolic elements. A standard uniformization (G, Δ) of a closed Riemann surface S is characterized topologically by the regular planar covering $h \circ \pi \colon \Delta \to S$.

Using a theorem on regular planar coverings from §2, we can describe this covering with the help of a finite set of disjoint planar loops $v_1, \ldots, v_m$ on S as follows: each of the loops $v_1, \ldots, v_m$ is lifted to a loop in Δ; the defining subgroup of the covering $\pi \colon \Delta \to S$ is the smallest normal subgroup of $\pi_1(S)$ spanned by the loops $v_1, \ldots, v_m$. It was mentioned also in §2 that any standard uniformization of a closed Riemann surface S can be constructed with the help of a finite set of loops $v_1, \ldots, v_m$ as indicated above. Combining these remarks, we get the following theorem, which in essence describes all the standard uniformizations:

THEOREM ON STANDARD UNIFORMIZATION. *Let S be a closed Riemann surface of genus $p > 0$, and let $\{v_1, \ldots, v_m\}$ be a set of simple disjoint loops on S. Then there exists a unique (to within conformal equivalence) standard uniformization (G, Δ) of S such that each factor subgroup of G is Fuchsian or elementary, and the covering $h \circ \pi \colon \Delta \to S$ is constructed from the smallest normal subgroup of $\pi_1(S)$ containing the loops $v_1, \ldots, v_m$.*

In this theorem two uniformizations (G_1, Δ_1) and (G_2, Δ_2) of S are said to be *conformally equivalent* if there exists a conformal homeomorphism $f \colon \Delta_1 \to \Delta_2$ such that the diagram

$$\begin{array}{ccc} \Delta_1 & \xrightarrow{f} & \Delta_2 \\ \pi_1 \downarrow & & \downarrow \pi_2 \\ \Delta_1/G_1 & & \Delta_2/G_2 \\ h_1 \searrow & & \swarrow h_2 \\ & S & \end{array}$$

commutes.

Suppose now that m is the smallest number of the loops indicated in the theorem. Then it is clear that $m \leq p$. This minimal set of loops divides S into surfaces $Y_1, \ldots, Y_s$, with each factor subgroup of G a covering subgroup for some Y_i, $i = 1, \ldots, s$.

Choose the loops $v_1, \ldots, v_m$ so that v_1 separates Y_2 and Y_1; v_2 separates Y_3 and $Y_1 \cup Y_2, \ldots$; v_{s-1} separates Y_s and $Y_1 \cup \cdots \cup Y_{s-1}$; and $v_1, \cdots, v_m$ are nonseparating loops lying entirely on the boundary of Y_s. Then deform the loops $v_1, \ldots, v_m$ in such a way that the preceding properties are preserved,

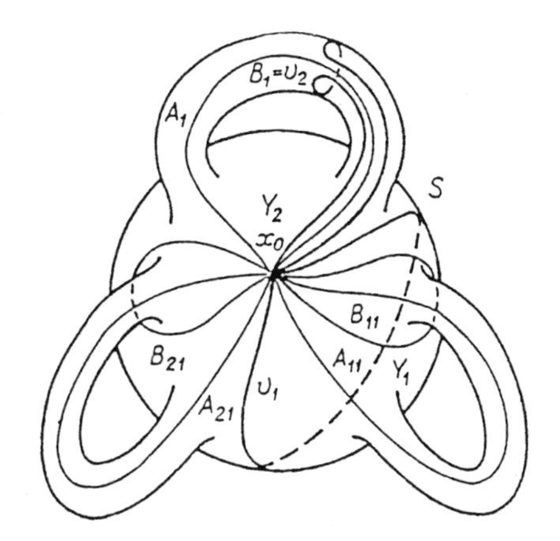

FIGURE 4

and $v_1, \ldots, v_m$ are tangent to one another at some point x_0. Then a system of generators of $\pi_1(S, x_0)$ can be chosen as follows. Let p_i be the genus of Y_i; it is clear that $\sum_1^s p_i \leq p$. If $p = \sum_1^s p_i + t$, then $t = m - s + 1$. For each $i = 1, \ldots, s$ there exist a canonical homotopy basis $\{A_{i1}, B_{i1}, \ldots, A_{ip_i}, B_{ip_i}\}$ for the surface Y_i with base point x_0, and loops $A_1, B_1, \ldots, A_t, B_t$, such that the family $\{A_{11}, \ldots, B_{1p_1}, \ldots, B_{sp_s}, A_1, \ldots, B_t\}$ is a canonical homotopy basis for S at x_0. In Figure 4 we indicate a canonical homotopy basis for a surface of genus 3 in the case when $s = 2$, $m = 2$, $p_1 = 1$, and $p_2 = 1$.

The following theorem describes the structure of the factor subgroups of a Kleinian group G giving a standard uniformization of a closed Riemann surface S.

(I). *Suppose that S is a closed surface of genus $p > 0$, $p = \sum_1^s p_i + t$ is a partition of p with $p_i > 0$, $i = 1, \ldots, s$ (or $s = 0$) and $t \geq 0$, and $\{A_{11}, \ldots, B_{1p_1}, A_{21}, \ldots, B_{sp_s}, A_1, \ldots, B_t\}$ is a canonical homotopy basis for S. Then there exists a unique standard uniformization (G, Δ) of S such that the following conditions are satisfied:*

a) *each factor subgroup of G is either elementary or Fuchsian;*

b) *for every $i = 1, \ldots, s$ the liftings of $A_{i1}, \ldots, B_{ip_i}$ beginning at a common base point generate a factor subgroup of G, and each factor subgroup of G is generated in this way;*

c) *each of the loops $A_1, \ldots, A_t$ can be lifted to a loop.*

Moreover, each standard uniformization is conformally equivalent to a uniformization obtained in this way.

This theorem yields an algebraic description of the Kleinian groups that give a standard uniformization:

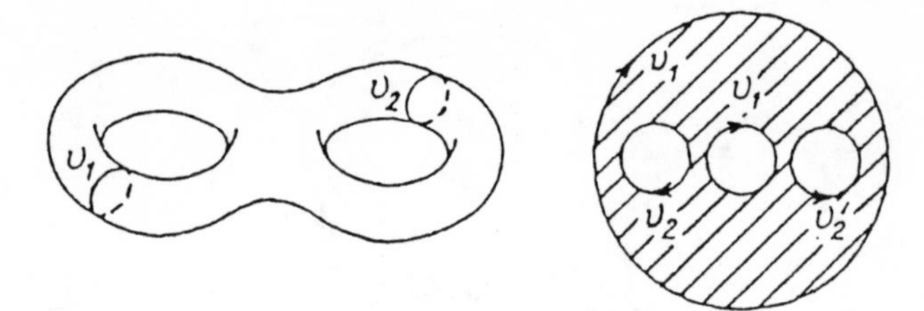

FIGURE 5

(II). *Suppose that a Kleinian group G gives a standard uniformization of a closed Riemann surface S. Then G has a presentation of the form*

$$G = \left\{ A_{11}, B_{11}, \dots, A_{sp_s}, B_{sp_s}, A_1, B_1, \dots, B_t; \right.$$

$$\left. \prod_j [A_{1j}, B_{1j}] = \cdots = \prod_j [A_{s_j}, B_{s_j}] = 1 \right\},$$

where $p = \sum_1^s p_i + t$, and either $p_i > 0$, $i = 1, \dots, s$, or $s = 0$.

If (G, Δ) is the standard uniformization of a Riemann surface S constructed in the theorem (II), then the Kleinian group G is called a *Koebe group*.

Let us consider some simple examples.

EXAMPLE 8. If $s = 1$ and $t = 0$, then there is only one factor subgroup, namely, the group G itself. In this case the uniformization is given by a Fuchsian group.

EXAMPLE 9. If $s = 0$ and $t = 2$, then there are two loops v_1 and v_2. These loops partition S into a region Y that is a sphere with four disks removed. Each connected component of $\pi^{-1}(Y)$ has a trivial stabilizer; therefore, all the factor subgroups of G are trivial and, consequently, G is a Schottky group (Figure 5).

EXAMPLE 10. If $s = 2$, $p_1 = p_2 = 1$, $m = 1$ and $t = 0$ (Figure 6), then there is a single loop v, which partitions S into two surfaces Y_1 and Y_2, and in this case G has many nontrivial factor subgroups. Each of the surfaces Y_1 and Y_2 is a torus with one disk removed, and thus each factor subgroup is a universal covering group of the torus. It is clear that G is the free product of two such groups.

We give a geometric picture of this case. Let G_1 be the group generated by the mappings $\gamma_1(z) = z + 4$ and $\gamma_2(z) = z + 4i$; a square serves as a

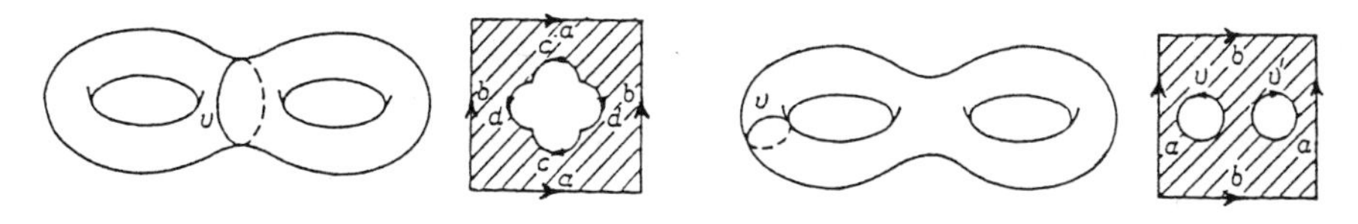

FIGURE 6 FIGURE 7

fundamental region F_1 of this group (Figure 6). The group G_2 obtained from G_1 by conjugation by the mapping $\gamma(z) = 1/z$ has a fundamental region F_2 which is the exterior of the quadrangle bounded by four circular arcs (Figure 6). The fundamental region $G = \langle G_1, G_2 \rangle$ is $F = F_1 \cap F_2$ (Figure 6).

EXAMPLE 11. If $s = 1$, $p_1 = 1$, $m = 1$ and $t = 1$ (Figure 7), then in this case there is only one loop v, which does not partition S. Then there exists only one surface Y, which is a torus with two disks removed. Clearly, all the factor subgroups of G are conjugate in G, and each of them is a universal covering group of a torus. The group G is a group of Schottky type, and a fundamental region of it is shown in Figure 7. Obviously, the subgroup generated by the mappings identifying the sides of the square is a factor subgroup.

These examples exhaust all the standard uniformizations of a surface of genus 2. The picture is the same for surfaces of greater genus, except that squares are replaced by $4p_i$-gons (the fundamental polygons of Fuchsian groups), and the fundamental regions can be bounded by a greater number of boundary components.

In conclusion, we mention that there are also more general uniformizations, for example, by Kleinian groups with torsion and with accidental parabolic elements. A description of them can be found in [186].

§5. Teichmüller spaces

We now pose the question of the extent to which the conformal structure of a Riemann surface of given genus p is uniquely determined. In his memoir on the theory of abelian functions (1857) Riemann observed that for $p > 1$ the classes of conformal equivalence of surfaces depend on $3p - 3$ complex parameters which he called *moduli*. Here he considered surfaces of algebraic functions (finite-sheeted branched coverings of the sphere); however, as noted in §1, this case actually encompasses all the closed Riemann surfaces.

The problem of studying the manifold of these parameters and its structure, known as the classical problem of moduli, was solved (completely, in a certain sense) only in the last decades. It had been noted already in the classical literature that the solution of the problem becomes simpler if conformal equivalence is replaced with a weaker equivalence by imposing certain topological restrictions, for example, by fixing generators of the fundamental group in the case of closed Riemann surfaces. This leads to a branched covering of a space of Riemann surfaces that can be described with the help of quasiconformal mappings. A program of investigations in this direction was sketched in papers of Teichmüller.

Let us begin by considering closed surfaces of genus $p > 1$. A *marking* of a surface S is defined to be the singling out of a system $\Sigma = \{a_1, b_1, \ldots, a_p, b_p \in \pi_1(S) \colon \prod_1^p [a_j, b_j] = 1\}$ of generators to within an inner automorphism of $\pi_1(S)$, and the pair (S, Σ) is called a *marked* Riemann surface. Since two homeomorphisms $S \to S'$ are homotopic if and only if the corresponding isomorphisms $\pi_1(S) \to \pi_1(S')$ coincide (to within inner automorphisms of the fundamental groups), it follows that the consideration of homeomorphisms of a marked surface (S, Σ) onto (S', Σ') is equivalent to taking a particular homotopy class of homeomorphisms $S \to S'$.

Moreover, *any* two marked surfaces of the same genus are homeomorphic and even quasiconformally equivalent. This follows from the topological equivalence of closed surfaces of the same genus and ***Nielsen's theorem*** that *every automorphism of the fundamental group $\pi_1(S)$ of a compact surface S can be realized geometrically, i.e., is induced by some topological automorphism of S* which is determined up to homotopy; with a homeomorphism $f \colon S \to S'$ at hand it is possible to construct a quasiconformal homeomorphism homotopic to it (for details see, for example, §6 in Chapter I of [130], where uniformization of surfaces is used in an essential way). This construction is generalized to arbitrary finitely generated groups and, in particular, to surfaces with finitely many deleted points.

Homeomorphisms $f' \colon S \to S'$ and $f'' \colon S \to S''$ are regarded as *equivalent* if $f'' \circ (f')^{-1}$ is homotopic to a conformal mapping $h \colon S' \to S''$. Passing to universal coverings of these surfaces (take the coverings to be the disk U), we get that f' and f'' can be lifted to automorphisms $\hat{f}'$ and $\hat{f}''$ of U such that the diagram

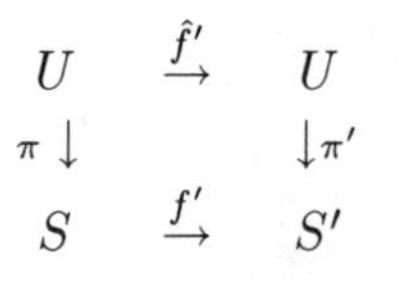

$$\begin{array}{ccc} U & \xrightarrow{\hat{f}'} & U \\ {\scriptstyle\pi}\downarrow & & \downarrow{\scriptstyle\pi'} \\ S & \xrightarrow{f'} & S' \end{array}$$

commutes (as does the analogous diagram for f''), and $\hat{f}'$ and $\hat{f}''$ coincide on the boundary (the circle) ∂U.

The set of equivalence classes $[S]$ of marked surfaces of genus p with n deleted points or, what is the same, the set of corresponding equivalence classes $[f]$ of quasiconformal homeomorphisms forms the *Teichmüller space $T(S)$ of the surface S*. In it we introduce the metric

$$d([f], [f']) = \frac{1}{2} \inf_{f \in [f], f' \in [f']} \ln K(f' \circ f^{-1}). \tag{35}$$

If, instead of S, we take another marked surface S_1, then we get an isometric isomorphism $T(S_1) \to T(S)$ by the formula $[f \circ h] \to [f]$, where h is a homeomorphism $S_1 \to S$. Therefore, one can speak of the simple space $T(p, n)$ (of equivalence classes) of surfaces of type (p, n). The isometries $[f \circ h] \to [f]$ generate the so-called *modular group* $\operatorname{Mod} T(p, n)$ of $T(p, n)$, which is the factor group of the full group of quasiconformal homeomorphisms of some surface S_0 by the normal subgroup of automorphisms homotopic to the identity; and the space $R(p, n)$ of Riemann surfaces themselves is obtained as the quotient $T(p, n)/\operatorname{Mod} T(p, n)$. It has been proved that $\operatorname{Mod} T(p, n)$ acts discontinuously in $T(p, n)$. For example, for tori ($p = 1$, $n = 0$) the space $T(1, 0)$ is conformally equivalent to the upper half-plane, and $\operatorname{Mod} T(1, 0)$ is isomorphic to the elliptic modular group.

The use of extremal quasiconformal mappings enables us to prove that $T(p, n)$ is a *cell* (i.e., homeomorphic to a Euclidean ball) of dimension $6p - 6 + 2n$ (see [13], [47], [127] and [229]). This is an old result obtained in his time by Fricke in a very complicated way (see [79]).

The Teichmüller space was and still is a subject of deep investigations by many authors. Here we are mainly interested in results relating to uniformization theory.

The main question in the theory of moduli is whether $T(p, n)$ admits a complex analytic structure. This question has also been investigated by many authors, who have introduced various local coordinates in $T(p, n)$. It has been proved that *$T(p, n)$ is a complex analytic manifold of dimension $m = 3p - 3 + n$*, and the complex structure compatible with the metric (35) is unique.

This result admits generalization to arbitrary Riemann surfaces, and we at once present the general case. We exclude the trivial cases $m = 0$ and $m = 1$ and consider surfaces of hyperbolic type.

Let S be a particular surface of this type. We represent it in the form $S = H/\Gamma$, where Γ is a Fuchsian group (without elliptic elements) acting in the upper half-plane H. Consider the complex Banach space $M^s(\mathbf{C}, \Gamma)$ of functions $\mu \in L_\infty(\mathbf{C})$ such that $\mu(z)\, d\bar{z}/dz$ is a Beltrami Γ-differential and $\mu(\bar{z}) = \overline{\mu(z)}$, with the norm $\|\mu\|_{M^s} = \|\mu\|_\infty$, and let $M_1^s(\mathbf{C}, \Gamma) = \{\mu \in M^s(\mathbf{C}, \Gamma)\colon \|\mu\| < 1\}$ be its unit ball. Let $Q(\Gamma)$ be the group (with respect to superposition) of quasiconformal automorphisms of H that are compatible with Γ and have fixed points $0, 1, \infty$,i.e., automorphisms f^μ with $\mu \in M_1^s(\mathbf{C}, \Gamma)$, and let $Q_0(\Gamma)$ be the normal subgroup of $Q(\Gamma)$ consisting of all the $f^\mu \in Q(\Gamma)$ which leave fixed all the points of the real axis $\mathbf{R}$. The factor group

$$T(\Gamma) = Q(\Gamma)/Q_0(\Gamma) \tag{36}$$

is called the *Teichmüller space of the group* Γ (and of the surface S).

The equivalence classes of the f^μ will be denoted by $[f^\mu]$. They again correspond to the homotopy classes of mappings of S.

In $Q(\Gamma)$ we introduce the Teichmüller metric

$$d(f^\mu, f^\nu) = \tfrac{1}{2} \ln K(f^\mu \circ (f^\nu)^{-1});$$

according to (36), this metric takes the form

$$d([f^\mu], [f^\nu]) = \frac{1}{2} \inf_{f \in [f^\mu \circ (f^\nu)^{-1}]} \ln K(f) \tag{35'}$$

in $T(\Gamma)$ and turns it into a complete connected metric space.

In particular, for $\Gamma = \mathbf{1}$ we obtain the *universal Teichmüller space* $T(\mathbf{1}) = T(H)$; all the $T(\Gamma)$ are imbedded in it in a natural way.

Suppose now that ω is an arbitrary homeomorphism in $Q(\Gamma)$ such that $\omega \Gamma \omega^{-1} = \Gamma$. It induces an isometric automorphism ω^* of the space $T(\Gamma)$ according to the formula $[f^\mu] \to [f^\mu \circ \omega^{-1}]$ and depends only on the values $\omega|_{\mathbf{R}}$; moreover, $(\omega_1 \circ \omega_2)^* = \omega_2^* \circ \omega_1^*$. The set of all such ω^* forms the *modular group* $\operatorname{Mod} T(\Gamma)$ *of the space* $T(\Gamma)$.

We now show that $T(\Gamma)$ *can be biholomorphically imbedded in a bounded region of a certain complex Banach space.* This result, due to Ahlfors [19] and Bers [48], is central in the theory of moduli and enables a global analytic structure on $T(\Gamma)$ to be defined.

Denote by $B(L, \Gamma)$ the Banach space of holomorphic functions $\varphi(z)$ in the lower half-plane $L = \{z\colon \operatorname{Im} z < 0\}$ such that the equation $\varphi(\gamma(z))\gamma'^2(z) = \varphi(z)$ holds for all $\gamma \in \Gamma$, with the norm $\|\varphi\| = \sup_L (\operatorname{Im} z)^2 |\varphi(z)|$. This is the space of automorphic forms of weight -4 that are bounded in the hyperbolic metric, or, in other words, the *quadratic differentials.* The dimension of $B(L, \Gamma)$ is finite if and only if Γ is a finitely generated group of the first kind.

Further, consider the space $M(H, \Gamma)$ of functions $\mu \in L_\infty(\mathbf{C})$ such that $\mu(z)\, d\bar{z}/dz$ is a Beltrami Γ-differential and $\mu(z) = 0$ in L, with the norm $\|\mu\|_M = \|\mu\|_\infty$. Let $M_1(H, \Gamma)$ be its unit ball, and let f_μ be the quasiconformal automorphism of $\overline{\mathbf{C}}$ with $\mu \in M_1(H, \Gamma)$ and with fixed points 0, 1, ∞; all such f_μ are conformal in L.

It is not hard to see that the equality $f_{\mu_1} = f_{\mu_2}$ holds on $\mathbf{R}$ if and only if $f^{\tilde{\mu}_1} = f^{\tilde{\mu}_2}$ on $\mathbf{R}$ for the corresponding $\tilde{\mu}_j \in M_1^s(\mathbf{C}, \Gamma)$ obtained from the μ_j by extension from H to L by symmetry; hence, there is a one-to-one correspondence between the restrictions $f_\mu|_{\mathbf{R}}$ and the classes $[f^{\tilde{\mu}}]$, i.e., the points of $T(\Gamma)$.

Indeed, $f^{\tilde{\mu}_j}(z) = h_j \circ f_{\mu_j}(z)$ for $z \in H$, where h_j is a conformal mapping of the region $f_{\mu_j}(H)$ onto H; thus, if $f^{\tilde{\mu}_1}$ and $f^{\tilde{\mu}_2}$ are equivalent, then the function

$$h(w) = \begin{cases} h_2 \circ h_1^{-1}(w), & w \in f_{\mu_1}(H), \\ f_{\mu_2} \circ f_{\mu_1}^{-1}(w), & w \in f_{\mu_1}(L), \end{cases}$$

extends to a quasiconformal automorphism of $\overline{\mathbf{C}}$ that is conformal in $\mathbf{C} \backslash f_{\mu_1}(\mathbf{R})$. Consequently, h is a Möbius mapping and, by the accepted normalization, reduces to the identity.

Thus, if instead of $[f^\mu]$ we start from the equivalence classes $[f_\mu]$, then we get (up to an isometric isomorphism) the same space $T(\Gamma)$.

The normalization of mappings adopted above is not essential. There is an invariant determining a locally univalent holomorphic function $f(z)$ in $D \subset \overline{\mathbf{C}}$ to within a Möbius transformation. This is the *Schwarzian derivative*

$$\{f, z\} = \left(\frac{f''(z)}{f'(z)}\right)' - \frac{1}{2}\left(\frac{f''(z)}{f'(z)}\right)^2, \qquad z = x + iy \in D. \tag{37}$$

The superposition $f_1 \circ f$ satisfies the relation

$$\{f_1 \circ f, z\} = (\{f_1, \varsigma\} \circ f) f'^2(z) + \{f, z\}, \qquad \varsigma = f(z), \tag{38}$$

which gives us that if $f_1 = \gamma$ is a Möbius transformation then $\{\gamma \circ f, z\} = \{f, z\}$, and for $f = \gamma$ the equality (38) can be written in the form

$$\{f_1 \circ \gamma, z\} = (\{f_1, \varsigma\} \circ \gamma)\gamma'^2(z).$$

But if $\varphi = \{f, z\}$ is given, then the function f itself has the form $f(z) = \eta_1(z)/\eta_2(z)$, where η_1 and η_2 are linearly independent solutions of the equation

$$2\eta''(z) + \varphi(z)\eta(z) = 0, \tag{39}$$

and from this it is clear that f is determined by φ to within a Möbius automorphism of $\overline{\mathbf{C}}$. We mention further that for $D = L$ we have the well-known estimate $|\{f, z\}| \le 3/2y^2$ (see [125] and [198]); and, as Nehari showed [198], the condition $|\{f, z\}| < 1/2y^2$ suffices for f to be univalent in L.

In particular, for $f(z) = f_\mu(z), z \in L$, the foregoing implies that $\varphi_\mu = \{f_\mu, z\} \in B(L, \Gamma)$, with $\|\varphi_\mu\| < 3/2$.

The derivatives of φ_μ are in a one-to-one correspondence with the classes $[f_\mu]$; thus, the correspondence $\kappa\colon [f_\mu] \to \varphi_\mu$ is a holomorphic homeomorphism $T(\Gamma) \to B(L, \Gamma)$ in the sense that if μ, as an element of $L_\infty(\mathbf{C})$, depends holomorphically on complex parameters then so does φ_μ, as an element of $B(L, \Gamma)$. (For clarification we mention that $T(\Gamma)$ can also be defined as the quotient $M_1(H, \Gamma)/M_0(H, \Gamma)$, where $M_0(H, \Gamma) = \{\mu \in M_1(H, \Gamma)\colon f_\mu|_{\mathbf{R}} = I\}$, with M_0 acting from the right in M_1 according to the superposition formula $f_{\mu\nu} \equiv f_\nu \circ f_\mu, \nu \in M_0, \mu \in M_1$.) Consequently, $\dim T(\Gamma) = \dim B(L, \Gamma)$ and, in particular, $\dim T(p, n) = 3p - 3 + n$.

It remains to show that the set $\kappa(T(\Gamma)) = \Delta(\Gamma)$ is open in $B(L, \Gamma)$ and, hence, is a region. When $\dim T(\Gamma) < \infty$ this follows immediately from the Brouwer theorem on invariance of domain; however, the proof is considerably more complicated in the general case. Ahlfors and Weill [23] modified Nehari's arguments [198] and proved that $\Delta(\Gamma)$ contains the ball $\|\varphi\|_{B(L,\Gamma)} < 1/2$, and, what is more, if for a given $\varphi \in B(L, \Gamma)$ with $\|\varphi\| < 1/2$ we set $\mu(z) = -2y^2\overline{\varphi(z)}$, $\operatorname{Im} z = y > 0$, then f_μ is compatible with Γ and $\{f_\mu, z\} = \varphi$. Together with (38) this leads to a proof that each $\varphi \in \Delta(\Gamma)$ is open; an independent proof is given in [48]. It is also known that when $\dim T(\Gamma) > 1$ the mapping $\mu \to \varphi_\mu$ can have only local but not global holomorphic sections. The quasi-Fuchsian groups $\Gamma_\mu = f_\mu \Gamma f_\mu^{-1}$ defining the surfaces $S_\mu = f_\mu(H)/\Gamma_\mu$ now serve as the images of Γ in $T(\Gamma)$.

In view of the results given above, the space $T(\Gamma)$ can be identified with its image in $B(L, \Gamma)$. Then the elements of $\operatorname{Mod} T(\Gamma)$ are biholomorphic automorphisms of $T(\Gamma)$, and $T(\Gamma)$ itself is $T(1) \cap B(L, \Gamma)$. This result has long been known for finite-dimensional spaces $T(\Gamma)$, and it was recently established in the general case by Tukia [305].

The *holomorpic convexity* of $T(\Gamma)$ is a very important property of this space. It was first established by Bers and Ehrenpreis [58] with use of the geometric criterion for quasicircles given in §4 of Chapter I. Krushkal′ [133] strengthened this result by showing that $T(\Gamma)$ *is a domain of bounded holomorphy.*

Let us now assume that H/Γ is a surface of finite type (p, n), i.e., $T(\Gamma) = T(p, n)$. Choosing in $B(L, \Gamma)$ a basis $\varphi_1, \dots, \varphi_m$ $(m = 3p - 3 + n)$ and taking the expansion $\varphi_\mu = \sum_1^m c_j \varphi_j$, we get a biholomorphic imersion of $T(p, n)$ in a bounded region in $\mathbf{C}^m$.

It turns out that the metric (35′) itself is completely determined by the complex structure of $T(p, n)$. Namely, as shown by Royden [213] (see also [72]), this metric coincides with the *Kobayashi metric* in $T(p, n)$ and, consequently, is biholomorphically invariant. Moreover, with a certain exception $\operatorname{Mod} T(p, n)$ is the full (countable!) group of biholomorphic automorphisms of $T(p, n)$; hence, $T(p, n)$ is trivially not a homogeneous region when $\dim T(p, n) > 1$.

As for the *Riemann spaces* $R(p, n) = T(p, n)/\operatorname{Mod} T(p, n)$, a general theorem of H. Cartan [63] gives us that this is a normal complex space. It has *nonuniformizable* singularities due (as Rauch [209] showed) to the presence of surfaces with nontrivial conformal automorphism groups.

The space $T(\Gamma)$ can be also described somewhat differently with the help of equation (39). Namely, we associate with *each* $\varphi \in B(L, \Gamma)$ the meromorphic function $w_\varphi(z) = \eta_1(z)/\eta_2(z)$ on L, where η_1 and η_2 are linearly independent solutions of (39) with the initial conditions $\eta_1(-i) = \eta_2'(-i) = 0$ and $\eta_2(-i) = \eta_1'(-i) = 1$. The function w_φ is locally univalent in L and has a decomposition $w_\varphi(z) = (z + i)^{-1} + O(|z + i|)$ near $z = -i$, while for each particular $z \in L \backslash \{-i\}$ the mapping $\varphi \to w_\varphi(z)$ defines a holomorphic function $B(L, \Gamma) \to \mathbf{C}$. The condition $\varphi(\gamma(z))\gamma'^2(z) = \varphi(z)$, $\gamma \in \Gamma$, implies the existence of a homomorphism $\chi_\varphi \colon \Gamma \to \mathcal{M}_2$ according to the formula $w_\varphi \circ \gamma = \chi_\varphi(\gamma) \circ w_\varphi$, and of the group $\Gamma_\varphi = \chi_\varphi(\Gamma)$ (called the *monodromy group* of φ); moreover, for each particular γ the elements $\chi_\varphi(\gamma)$ together with w_φ depend holomorphically on $\varphi \in B(L, \Gamma)$.

The elements $\varphi \in T(\Gamma)$ *are characterized* by the fact that the Γ_φ are *quasi-Fuchsian* groups (coinciding with Γ_μ to within conjugation in $\mathcal{M}_2$) with invariant quasicircles $w_\varphi(\overline{\mathbf{R}})$ (so that $w_\varphi(L)$ is a Jordan region) and w_φ can be extended from L to a quasiconformal automorphism $W_\varphi \colon \overline{\mathbf{C}} \to \overline{\mathbf{C}}$ that is compatible with Γ_φ. Let $D_\varphi = \overline{\mathbf{C}} \backslash \overline{w_\varphi(L)}$.

If Γ is of finite type, then the $\varphi \in \partial T(\Gamma)$ correspond to the *b-groups* with invariant components $w_\varphi(L)$ (on this, see §5 in Chapter I). For $\varphi \in B(L,\Gamma)\backslash\overline{T(\Gamma)}$ the groups Γ_φ are not discrete in general.

Let us now pass directly to uniformization questions. With this aim we introduce a *fiber bundle* $\tilde{T}(\Gamma)$ over $T(\Gamma)$ by setting (see, for example, [54])

$$\tilde{T}(\Gamma) = \{(\varphi, z) \colon \varphi \in T(\Gamma), z \in D_\varphi\}.$$

This is a bounded region in $B(L,\Gamma)\times\mathbf{C}$. The space $\tilde{T}(\Gamma)$ has a natural complex structure, determined from the requirement that the projection $(\varphi, z) \to \varphi$ be holomorphic; the fibers over the points $\varphi \in T(\Gamma)$ are the universal coverings D_φ of the surfaces $S_\varphi = D_\varphi/\Gamma_\varphi$.

The action of Γ can be extended holomorphically to the whole of $\tilde{T}(\Gamma)$ by the formula $\gamma(\varphi, z) = (\varphi, W_\varphi \circ \gamma \circ W_\varphi^{-1})$, $\gamma \in \Gamma$; then $\tilde{T}(\Gamma)/\Gamma$ is the bundle over $T(\Gamma)$ with fibers S_φ, and these fibers depend holomorphically on φ. Thus, the set of surfaces S_φ is turned into a *holomorphic family* of Riemann surfaces, with the conformal structure on each of them induced by the complex structure of $\tilde{T}(\Gamma)/\Gamma$ for a particular φ.

The problem of *simultaneous uniformization of all differentials* of a given order on different surfaces can be solved naturally in terms of the space $\tilde{T}(\Gamma)$. We present the corresponding result of Bers [48] (see also Chapter V in [130]).

For an integer $q \geq 2$, denote by $B_q(D_\varphi, \Gamma_\varphi)$ the space of holomorphic Γ_φ-automorphic forms of weight $-2q$ on D_φ, i.e., the space of solutions of the equation $\psi(\gamma(z))\gamma'^q(z) = \psi(z), \gamma \in \Gamma$, that are holomorphic in D_φ, with the norm $\|\psi\| = \sup_{D_\varphi} \lambda_\varphi^{-q}(z)|\psi(z)|$, where $\lambda_\varphi(z)|dz|$ is the hyperbolic metric in D_φ.

Each form $\psi \in B_q(H,\Gamma)$ extends canonically to a holomorphic function $\Psi(\varphi, z)$ on $\tilde{T}(\Gamma)$ with $\Psi(0,z) = \psi(z)$, which establishes for each $\varphi \in T(\Gamma)$ a linear topological isomorphism of $B_q(H,\Gamma)$ onto $B_q(D_\varphi, \Gamma_\varphi)$ according to the formula $\psi \to \Psi(\varphi, z)$. This function has the form

$$\Psi(\varphi, z) = -\frac{2q+1}{\pi} \iint\limits_{\eta<0} \frac{\eta^{2q-2}\psi(\bar{\varsigma})w_\varphi'(\varsigma)^q \, d\xi \, d\eta}{(w_\varphi(\varsigma) - z)^{2q}}.$$

By suitably choosing Ψ, it is possible to give a parametric representation of *all* the algebraic curves of given genus $p > 1$ (by analogy with the fact that the function $\wp(z;1,\tau)$ and its derivative $\wp'(z;1,\tau)$ uniformize all curves of genus 1).

We mention further that, by using the bundle over $T(\Gamma)$, Griffiths [90] succeeded in uniformizing certain subsets of projective algebraic varieties of arbitrary dimension.

Certain geometric properties of Riemann surfaces and Teichmüller spaces are given in Problem 67 in §4 of Chapter IV.

We now introduce spaces of deformations of hyperbolic and conformal structures (see §8) on arbitrary manifolds of dimension $n \geq 2$. It will be seen that these spaces coincide for $n = 2$; but that is no longer the case in dimensions $n \geq 3$.

Let M be an arbitrary n-dimensional manifold, and denote by $H(M)$ the set of pairs (N, f), where N is a complete hyperbolic manifold, and $f\colon M \to N$ is a homotopy equivalence. We introduce on $H(M)$ the following equivalence relation: a pair (N_1, f_1) is said to be *equivalent* to a pair (N_2, f_2) if there exists a isometry $\varphi\colon N_1 \to N_2$ such that the following diagram is commutative:

$$\begin{array}{ccc} M & \xrightarrow{f_1} & N_1 \\ \downarrow \mathrm{id} & & \downarrow \varphi \\ M & \xrightarrow{f_2} & N_2. \end{array}$$

The quotient space of $H(M)$ with respect to this equivalence is called the space of deformations of hyperbolic structures on the manifold M, or the *Teichmüller space of hyperbolic structures on* M, and is denoted by $T_H(M)$. Topologies can be introduced on $T_H(M)$ in various ways. We describe two of them. The first is the so-called *algebraic topology.* It is introduced as follows. The set $H(M)$ can be identified with the set of conjugacy classes of faithful discrete representations of the fundamental group of M in the group of isometries of the n-dimensional hyperbolic space H^n. The space of representations has an obvious topology, namely, the topology of convergence on generators. Then on $T_H(M)$ we introduce the quotient topology of the space of representations modulo the action of the group $\operatorname{Isom} H^n$ by conjugations. The space $T_H(M)$ with this topology is denoted by $T_H^A(M)$.

Another topology—the *quasi-isometric topology*—is introduced as follows. Namely, a small neighborhood of a point $[N, f] \in T_H(M)$ consists of all the points $[N', f'] \in T_H(M)$ such that there exists a diffeomorphism $g\colon N \to N'$ for which $g \circ f$ is homotopic to f', and g is a quasi-isometry with small quasi-prometric coefficient. The space $T_H(M)$ with this topology is denoted by $T_H^Q(M)$. We remark that if $n = 2$, then it can be proved that $T_H^Q(M)$ is naturally isomorphic to the Teichmüller space $T(M)$ introduced earlier for the hyperbolic surface M.

Suppose now that $n \geq 3$ and M is a complete hyperbolic manifold of finite volume. Then $T_H^Q(M)$ consists of a single point. This is a consequence of the following rigidity theorem:

THEOREM (MOSTOW [194]). *Let M and M' be n-dimensional, $n \geq 3$, complete hyperbolic manifolds of finite volume, and $f\colon M \to M'$ a homotopy equivalence. Then M and M' are isometric, and the isometry can be chosen in the homotopy class of the mapping f.*

The class of manifolds for which the rigidity theorem holds does not consist only of manifolds of finite volume. Recent results of Sullivan [300] imply that if $M = H^n/\Gamma$, where Γ is a group of divergent type, then the rigidity theorem holds for M, and, consequently, $T_H^Q(M)$ consists of a single point for such manifolds.

Now let $K(M)$ be the set of pairs (N, f), where N is a conformal manifold, and $f: M \to N$ is a homotopy equivalence. A pair (N_1, f_1) will be said to be equivalent to a pair (N_2, f_2) if there exists a conformal homeomorphism $\varphi: N_1 \to N_2$ such that the diagram

$$\begin{array}{ccc} M & \xrightarrow{f_1} & N_1 \\ \downarrow \mathrm{id} & & \downarrow \varphi \\ M & \xrightarrow{f_2} & N_2 \end{array}$$

is commutative.

The quotient space $K(M)$ with respect to this equivalence is denoted by $T_K(M)$ and is called the space of deformations of conformal structures on M, or the *Teichmüller space of conformal structures on M*.

We first introduce an algebraic topology on $T_K(M)$.

If N is an n-dimensional conformal manifold, then a homomorphism of the holonomy of the conformal structure on N determines a representation (not necessarily faithful) of the fundamental group of M in the Möbius group $\mathcal{M}(n)$. Therefore, the quotient topology of the space of such representations modulo the action of the group $\mathcal{M}(n)$ by conjugations, can be introduced by $T_K(M)$ in a way analogous to that above. The space $T_K(M)$ with this topology is denoted by $T_K^A(M)$.

We now introduce the so-called *quasiconformal topology* on $T_K(M)$. Namely, a small neighborhood of a point $[N, f] \in T_K(M)$ is taken to be the set of all points $[N', f'] \in T_K(M)$ for which there exists a diffeomorphism $g: N \to N'$ such that $g \circ f$ is homotopic to f', and g is quasiconformal with quasiconformality coefficient close to 1. The space $T_K(M)$ with this topology is denoted by $T_K^K(M)$. It can be shown that if $n = 2$, then $T_K^K(M)$ is naturally isomorphic to the Teichmüller space $T(M)$ introduced above for the surface M.

Thus, the spaces $T(M)$, $T_H^Q(M)$, and $T_K^K(M)$ coincide in a definite sense for $n = 2$.

For $n \geq 3$ the spaces $T_H^Q(M)$ and $T_K^K(M)$, where M is a fixed hyperbolic manifold, are no longer isomorphic, not even when M is closed. Examples illustrating this can be obtained from results of Apanasov [36]. Sullivan [300] recently used ideas of Thurston to show that for most hyperbolic manifolds M the space $T_K^K(M)$ is nontrivial; for this to be so it suffices that M contain a geodesic submanifold of dimension $n - 1$.

The algebraic topology on the spaces $T_H(M)$ and $T_K(M)$ is exploited to study the spaces of deformations of Kleinian groups both on the plane and in space. While the space $T_K^A(M)$ has been fairly thoroughly studied for $n = 2$, its topological properties are practically unknown in larger dimensions.

§6. Branched coverings of Riemann surfaces

As mentioned in §2, the construction of coverings $\tilde{X}$ of a given surface X (from corresponding subgroups of its fundamental group) leads to *unbranched* coverings. However, the branched coverings have a number of specific properties which we shall consider here.

A holomorphic mapping $\pi: X \to Y$ of a Riemann surface X onto a Riemann surface Y is called an *unbounded covering* if the following two conditions are met:

(a) π^{-1} is a locally proper mapping, i.e., each point $y \in Y$ has an open neighborhood U_y such that the restriction of π to each connected component of $\pi^{-1}(U_y)$ is a proper mapping onto U_y.

(b) There exists a discrete subset $Y_0 \subset Y$ such that the restriction π: $X \backslash \pi^{-1}(Y_0) \to Y \backslash Y_0$ is a covering.

The set Y_0 is called the *branching set* of the covering $\pi: X \to Y$. In the case $Y_0 = \varnothing$ we obtain the unbranched coverings whose main properties were presented in §2.

Any point $y_0 \in Y_0$ has a simply connected neighborhood V such that $V \cap Y_0 = \{y_0\}$ and $\pi^{-1}(V)$ consists of connected components U_i $(i \in I)$, each carried into V by the proper mapping $\pi|_{U_i}$. Moreover, for each component U_i the cardinality of the set $E_i = \pi^{-1}(y_{0*}) \cap U_i$ $(y_{0*} \in V\backslash\{y_0\})$ is finite and does not depend on the choice of the point y_{0*}. Denote it by $\beta_\pi(y_i)$ and call it the *branching order* of the mapping at the point y_i. For $y \in Y \backslash Y_0$ let $\beta_\pi(y) = 1$.

A branched covering $\pi: \tilde{S} \to S$ of a Riemann surface S is said to be *regular* if the group $G(\tilde{S}, S)$ of covering homeomorphisms of $\tilde{S}$ acts transitively on the fibers of π^{-1}, i.e., for any pair of points $x, y \in \tilde{S}$ such that $\pi(x) = \pi(y)$ there is an element $h \in G(\tilde{S}, S)$ such that $h(x) = y$.

A branched covering $\pi: \tilde{S} \to S$ is called a *covering of regular type* if the function β_π is constant on each fiber of π^{-1}, i.e., $\beta_\pi(x) = \beta_\pi(y)$ for any $x, y \in \tilde{S}$ with $\pi(x) = \pi(y)$.

Every regular covering is a covering of regular type. The converse is not true in general (see Example 13). However, Greenberg showed [89] that if $\tilde{S}$ is simply connected, then every covering of regular type is regular.

Let X be a Riemann surface, Γ a properly discontinuous group of conformal automorphisms of X, and Γ_0 a subgroup of Γ. The mapping $\pi: X/\Gamma_0 \to X/\Gamma$ acting by the rule $\pi(\Gamma_0 x) = \Gamma x, x \in X$, defines a covering of the Riemann surface X/Γ_0 over the Riemann surface X/Γ and is called *the covering induced by the inclusion* $\Gamma_0 \subset \Gamma$.

If Γ_0 and Γ are groups with signature that act on X such that $\Gamma_0 \subset \Gamma$ and Γ_0 does not contain elements of finite order, then the covering $\pi: X/\Gamma_0 \to X/\Gamma$ induced by the inclusion is a covering of regular type. Conversely, for any covering $\pi: X \to Y$ of regular type there exists a universal covering $\tilde{\pi}: \tilde{X} \to X$ such that π is induced by the groups $\Gamma_0 = G(\tilde{X}, X)$ and $\Gamma = G(\tilde{X}, Y)$

acting on $\tilde{X}$. Moreover, Γ does not contain elements of finite order, and $G(X, Y) \cong N(\Gamma_0, \Gamma)/\Gamma_0$, where $N(\Gamma_0, \Gamma)$ is the normalizer of the group Γ_0 in Γ.

In particular, knowledge of the group Γ of covering transformations of the universal covering $\tilde{\pi}$ enables us to determine the group $\operatorname{Aut} X$ of all conformal automorphisms of the surface X. Namely, we have a canonical isomorphism $\operatorname{Aut} X \cong N(\Gamma, \operatorname{Aut} \tilde{X})/\Gamma$.

Let Γ be a Fuchsian group with signature $(g, r; m_1, \dots, m_r)$ in the disk U, and let Γ_0 be a subgroup of Γ of finite index. The multiplicity of the covering $\pi\colon U/\Gamma_0 \to U/\Gamma$ induced by the inclusion $\Gamma_0 \subset \Gamma$ coincides with the index $|\Gamma : \Gamma_0|$ of Γ_0 in Γ and can be determined by the classical ***Riemann-Hurwitz formula***(8)

$$|\Gamma : \Gamma_0| = \mu(\Gamma_0)/\mu(\Gamma), \tag{41}$$

where $\mu(\Gamma) = 2g - 2 + \sum_{j=1}^{r}(1 - 1/m_j)$ is the hyperbolic volume of a fundamental set of Γ.

As noted by MacLachlan [162], formula (41) remains in force also for finitely generated Fuchsian groups of the second kind, but then the ratio on the right-hand side loses its geometric meaning, since the hyperbolic area of a fundamental region becomes infinite.

We illustrate the foregoing by examples.

EXAMPLE 12. Let us construct the covering induced by the action of a cyclic group Γ on the plane $\overline{\mathbf{C}}$. Suppose that Γ is generated by an elliptic element γ of order p. Then it can be presented in the form $\Gamma = \{\gamma, \gamma^{-1} : \gamma^p = (\gamma^{-1})^p = \gamma\gamma^{-1} = 1\}$. Choose γ to be the Möbius transformation $z' = \gamma(z)$ with fixed points $\pm i$ and such that

$$(z' + i)/(z' - i) = e^{2\pi i/p}(z + i)/(z - i).$$

Let I and I' be congruent arcs of circles whose endpoints meet in pairs at the points $\pm i$ at an angle of $2\pi/p$ (Figure 8). Denote by F the lune in $\overline{\mathbf{C}}$ bounded by the Jordan curve $I \cup I'$ and containing the point at infinity j it is a fundamental polygon for Γ. The transformation γ carries the side I into I'. After identification of these sides the lune F becomes a compact Riemann surface of genus 0. The canonical projection $\pi\colon \overline{\mathbf{C}} \to \overline{\mathbf{C}}/\Gamma$ gives a p-sheeted covering of a sphere over a sphere and has branching order p at the points $\pm i$. The covering π is not branched at the remaining points of $\overline{\mathbf{C}}$. The branching diagram of π is given in Figure 8.

EXAMPLE 13. Consider a group Γ with signature $(0, 3; 2, 3, 6)$ acting on $\overline{\mathbf{C}}$. It can be presented in the form

$$\Gamma = \{\gamma_1, \gamma_2, \gamma_3 : \gamma_1^2 = \gamma_2^3 = \gamma_3^6 = \gamma_1\gamma_2\gamma_3 = \mathbf{1}\}.$$

---

(8) *Translation editor's note*: Formula (40) has been eliminated by one of the authors in a revision.

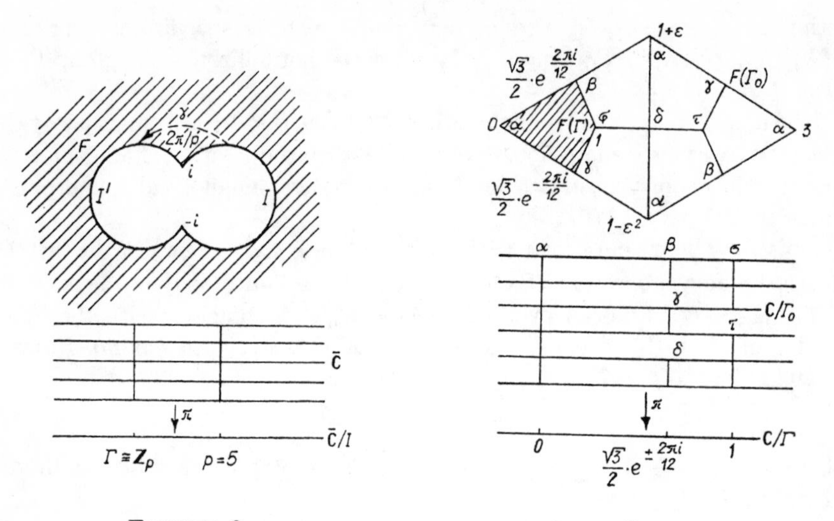

FIGURE 8 FIGURE 9

The following Möbius transformations can be taken as generators of Γ (see [285, Theorem 1.6]):

$$\gamma_1(z) = -(z-(1+\varepsilon^5)/2)+(1+\varepsilon^5)/2, \gamma_2(z) = \varepsilon^2(z-1)+1 \text{ and } \gamma_3(z) = \varepsilon z,$$

where $\varepsilon = e^{2\pi i/6} = 1/2 + i\sqrt{3}/2$.

Using the Reidemeister-Schreier method (see [164]), we can single out a subgroup Γ_0 of Γ of index 6 that does not contain elements of finite order; $g_1(z) = z+1+\varepsilon$ and $g_2(z) = z+1-\varepsilon^2$ can be taken as generators of Γ_0. Moreover, Γ_0 has the presentation $\Gamma_0 = \{g_1, g_2 \colon g_1 g_2 g_1^{-1} g_2^{-1} = 1\}$ and is isomorphic to the fundamental group of a torus.

The quadrangle $F(\Gamma)$ with vertices $0, (\sqrt{3}/2)e^{2\pi i/12}, 1$ and $(\sqrt{3}/2)e^{-2\pi i/12}$ serves as a fundamental set for the group Γ. It consists of two right triangles with angles $\pi/6, \pi/2$, and $\pi/3$, obtained one from the other by reflection with respect to the hypotenuse lying on the real axis. Further, the congruent legs of the triangles are identified under the action of Γ, and turn $F(\Gamma)$ into a compact Riemann surface of genus 0 with the triple of distinguished points $0, 1$, and $(\sqrt{3}/2)e^{2\pi i/12} \equiv (\sqrt{3}/2)e^{-2\pi i/12}$ over which $\Phi \colon \mathbf{C} \to \mathbf{C}/\Gamma$ has branching orders 6, 3, and 2, respectively.

The parallelogram $F(\Gamma_0)$ with vertices $0, 1+\varepsilon, 3, 1-\varepsilon^2$ serves as the fundamental polygon for Γ_0; under the action of Γ it is covered by six copies of $F(\Gamma)$. Identifying the opposite sides of $F(\Gamma_0)$, we obtain a torus $\mathbf{C}/\Gamma_0$. The covering $\pi \colon \mathbf{C}/\Gamma_0 \to \mathbf{C}/\Gamma$ of the torus over a sphere is branched at precisely six points, denoted in Figure 9 by Greek letters, and the branching orders can be determined from the diagram in Figure 9. The covering π is regular, and its group of covering transformations is cyclic of order six.

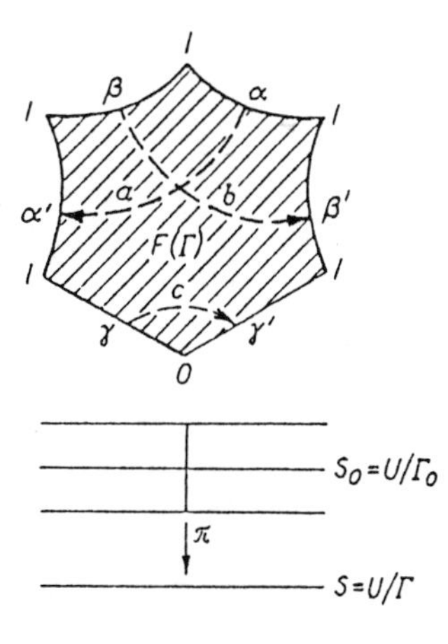

FIGURE 10

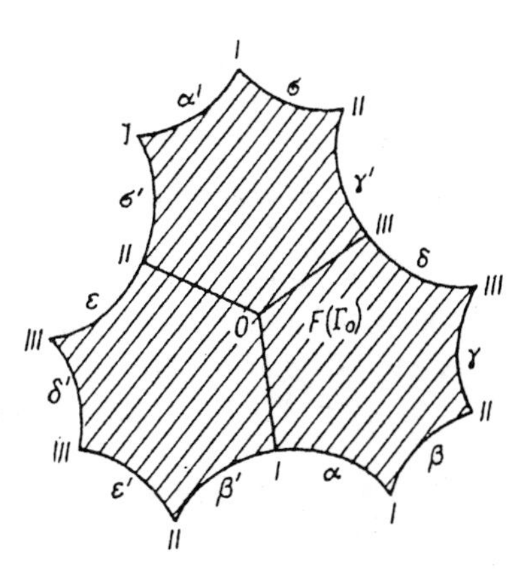

FIGURE 11

EXAMPLE 14. *A covering of regular type with trivial group of covering transformations.* Let Γ be a Fuchsian group in the unit disk U, with presentation $\Gamma = \{a, b, c : c^3 = c[a, b] = \mathbf{1}\}$. The non-Euclidean hexagon pictured in Figure 10 can be taken as the fundamental polygon $F(\Gamma)$ of this group; the elliptic vertex of $F(\Gamma)$ with angle $2\pi/3$ is at the origin.

We define an epimorphism θ of Γ onto the symmetric group S_3 according to the rule $\theta(a) = (1\ 2), \theta(b) = (2\ 3)$ and $\theta(c) = (1\ 3\ 2)$. The subgroup $G_0 \subset S_3$ leaving fixed the symbol 3 consists of two elements: $G_0 = \{(1), (1\,2)\}$. The group $\Gamma_0 = \theta^{-1}(G_0)$ has the following complete system of generators: $a, cac, c^2ac^2, bc^2, cb, c^2bc$.

Consider the non-Euclidean polygon

$$F = F(\Gamma) \cup c(F(\Gamma)) \cup c^2(F(\Gamma)). \tag{42}$$

Its sides are identified pairwise by the generators of Γ_0 as indicated in Figure 11. By the Poincaré theorem, the polygon (42) is a fundamental polygon of the group Γ_0 in U.

The canonical mapping $U \to U/\Gamma$ uniformizes the torus $S = U/\Gamma$ with a single distinguished point over which all branching orders are equal to 3. It is obtained from the polygon $F(\Gamma)$ as indicated in Figure 10. Here the elliptic vertex at the origin corresponds to the distinguished point on the torus.

The three-sheeted covering $\pi \colon U/\Gamma_0 \to U/\Gamma$ is branched over the same distinguished point and has order 3 over it, because the covering $U \to U/\Gamma_0 = S_0$ is unbranched. The covering diagram for π is pictured in Figure 10.

Let us compute the group $G(S_0, S)$ of covering transformations. We have that

$$G(S_0, S) \cong N(\Gamma_0, \Gamma)/\Gamma_0 \cong N(G_0, S_3)/G_0 \cong \mathbf{1}. \tag{43}$$

This chain of isomorphisms (43) shows that the transformation group $G(S_0, S)$ is trivial.

Other examples of coverings of regular type with trivial group of covering transformations can be found in Lemma 2 in [189].

§7. Methods of three-dimensional topology

1. *Kleinian groups and three-dimensional manifolds.* General properties of Kleinian groups in $\mathbf{R}^n$ $(n \geq 3)$ were studied in §2 of Chapter I. Here we dwell in more detail on groups acting in $\mathbf{R}^3_+$ (and discontinuous in $\overline{\mathbf{R}}^2 = \overline{\mathbf{C}}$). They are closely connected with three-dimensional manifolds, and this enables us to apply methods of three-dimensional topology to the investigation of the properties of Kleinian groups on the plane. Difficult problems in the planar theory of Kleinian groups have been solved in this way (which was developed mainly by Marden [168]). These questions are of interest also for topology itself, since the three-dimensional manifolds that can be uniformized by Kleinian groups have been little studied.

Accordingly, let $G \subset \mathcal{M}_2$ be a Kleinian group, extended to $\mathbf{R}^3_+$ (see §1 in Chapter I) and acting discontinuously on the set $\mathbf{R}^3_+ \cup \Omega(G)$. The three-dimensional manifold $M(G) = (\mathbf{R}^3_+ \cup \Omega(G))/G$ is connected in the natural way with the group G. This manifold is orientable and has boundary $\partial M(G) = \Omega(G)/G$ which is a union of Riemann surfaces.

Let $\pi \colon \mathbf{R}^3_+ \cup \Omega(G) \to M(G)$ be the canonical projection. The mapping π is a covering (branched, in general). But if G does not contain elliptic elements, then π is not branched, and $\mathbf{R}^3_+ \cup \Omega(G)$ is the universal covering manifold for $M(G)$. In this case the fundamental group $\pi_1(M(G))$ is isomorphic to G.

It is assumed everywhere below that the Kleinian groups under consideration do not contain elliptic elements. Then the manifold $M(G)$ can be studied with the help of known methods in three-dimensional topology (see, for example, the articles [238] and [239] of Waldhausen).

We give some properties of manifolds $M(G)$ that are uniformizable by Kleinian groups.

1) $M(G)$ is *irreducible*; this means that each two-dimensional sphere imbedded in $M(G)$ bounds a ball in $M(G)$.

2) $M(G)$ is *aspherical* in dimensions 2 and 3, i.e., $\pi_2(M(G)) = \pi_3(M(G)) = 0$.

3) $M(G)$ is a *Poincaré manifold*, i.e., each homotopy sphere imbedded in $M(G)$ is homeomorphic to a Euclidean sphere.

4) If G is a nonelementary group, then $\pi_1(M(G))$ has trivial center, and $\pi_1(S)$ is nonabelian for each component S of the boundary of $M(G)$.

5) $M(G)$ is *sufficiently large* in the sense of Waldhausen, i.e., $M(G)$ contains an incompressible surface (see Problem 118).

We take M to be an arbitrary three-dimensional manifold below. A *loop* on M is defined to be a continuous mapping of the circle S^1 into M; a *simple loop* is a homeomorphic mapping of S^1 into M.

The following theorems are fundamental in the investigation of three-dimensional manifolds.

LOOP THEOREM [100]. *If l is a nontrivial loop on ∂M (l represents a nontrivial element of the group $\pi_1(\partial M)$) that is trivial in $M(G)$, then there exists a simple loop l' that is nontrivial on $\partial M(G)$ and also trivial in $M(G)$.*

DEHN'S LEMMA [100]. *Let l be a simple nontrivial loop on $\partial M(G)$ that is trivial in $M(G)$. Then there exists a topological disk D in $M(G)$ such that $l = \partial D$.*

CYLINDER THEOREM [100]. *Let l_1 and l_2 be disjoint loops on $\partial M(G)$, neither contractible to a point in $M(G)$, and suppose that l_1 is freely homotopic to l_2 in $M(G)$. Then there exist simple loops l_1' and l_2', nontrivial in $M(G)$, that are boundary components of a topological cylinder in $M(G)$.*

VAN KAMPEN'S THEOREM [100]. *Let D be a topological disk in $M(G)$ whose boundary belongs to $\partial M(G)$. If $M(G)\backslash D$ has two components M_1 and M_2, then $\pi_1(M(G))$ is isomorphic to the free product of $\pi_1(M_1)$ and $\pi_1(M_2)$. But if $M(G)\backslash D$ is connected, then $\pi_1(M(G))$ is isomorphic to the free product of $\pi_1(M(G)\backslash D)$ and an infinite cyclic group.*

KNESER'S THEOREM [100]. *Let M be a compact three-dimensional manifold such that each component $S \subset \partial M$ is incompressible in M, i.e., the homomorphism $\pi_1(S) \to \pi_1(M)$ induced by the imbedding $S \subset M$ is a monomorphism. If $\pi_1(M)$ is the nontrivial free product of groups G_1 and G_2, then there exists a two-dimensional sphere imbedded in M that makes M into a connected sum of manifolds M_1 and M_2 (see subsection 2 below) with $\pi_1(M_i)$ isomorphic to G_i $(i = 1, 2)$, and $\pi_1(M)$ is isomorphic to the free product $\pi_1(M_1) * \pi_1(M_2)$.*

SPHERE THEOREM [100]. *If M is an orientable three-dimensional manifold such that $\pi_2(M) \neq 0$, then there exists a two-dimensional sphere imbedded in M that represents a nontrivial element of $\pi_2(M)$.*

The fundamental polyhedron $F(G)$ of the group G in $\mathbf{R}^3_+$ was already defined in the preceding chapter. It is frequently more convenient to consider the polyhedron $\hat{F}(G) = \overline{F}(G) \cap (\mathbf{R}^3_+ \cup \Omega(G))$. Then $M(G)$ can be obtained from $\hat{F}(G)$ by identifying equivalent faces.

We dwell further on the connection between parabolic subgroups of G and submanifolds of $M(G)$. Let $p \in \Lambda(G)$ be a fixed point of a parabolic element of G. Denote by G_p the stabilizer of p in G. It is known that G_p consists solely of the parabolic elements with the common fixed point and is either a cyclic group or a free abelian group of rank 2. If G_p is cyclic, then $M(G)$ contains a submanifold homeomorphic to $\{z \in \mathbf{C}: 0 < |z| < 1\} \times (0,1)$, which is called an *open punctured cylinder*; but if G_p is a free abelian group of rank 2, then $M(G)$ contains a submanifold homeomorphic to $\{z \in \mathbf{C}: 0 < |z| < 1\} \times S^1$ (S^1 is the circle), which is called an *open punctured torus.*

Marden showed [168] that if G is geometrically finite, then $M(G)$ is the union of a compact submanifold of $M(G)$ and finitely many open punctured tori and cylinders, and conversely.

Elementary manifolds uniformizable by Kleinian groups are described in Examples 54–57.

Using methods of three-dimensional topology, Marden made considerable progress in the Poincaré-Ahlfors program for investigating Kleinian groups in $\mathbf{R}^3_+$.

By the same methods, Gusevskiĭ [92] gave a description of the finitely generated Kleinian groups with connected set of discontinuity, and of the Kleinian groups isomorphic to the fundamental group of a closed nonorientable surface. Moreover, he proved that the doubly generated nonelementary torsion-free Kleinian groups are free, and he gave a complete classification of such groups with finite-sided fundamental polyhedron.

2. ***Construction of three-dimensional manifolds.*** In this subsection we describe the main ways of constructing three-dimensional manifolds. Some of them admit realization in terms of the theory of Kleinian groups and are closely connected with the combination method, the structure of fundamental polyhedra, and so on. However, mainly these are purely topological methods.

a) *Connected sums.* Let M, M_1, and M_2 be three-dimensional manifolds (with or without boundary). Assume that there exist three-dimensional topological balls $B_i \subset \operatorname{int} M_i$ $(i = 1, 2)$ and imbeddings $h_i: S_i \to M$ $(S_i = M_i \setminus \operatorname{int} B_i)$ such that $h_1(S_1) \cap h_2(S_2) = h_1(\partial B_1) = h_2(\partial B_2)$ and $M = h_1(S_1) \cup h_2(S_2)$. Then M is called a *connected sum* of M_1 and M_2, denoted by $M = M_1 \# M_2$. If also M, M_1, and M_2 are orientable, then it is required that the imbeddings $h_i: S_i \to M$ $(i = 1, 2)$ preserve orientation.

The construction of three-dimensional manifolds with the help of the connected sum can be realized by spatial Klein combination. This is shown by Example 83.

b) *Disk sums.* Let M, M_1, and M_2 be three-dimensional manifolds with boundary, let D_i be a topological disk in the boundary of M_i $(i = 1, 2)$, and let f be a homeomorphism mapping D_1 onto D_2. The manifold M obtained by gluing M_1 and M_2 according to the mapping f is called a *disk sum* of M_1 and M_2, denoted by $M = M_1 \triangle M_2$. If M, M_1, and M_2 are orientable, then it is required that f preserve orientation.

The manifold $M(G) = (\mathbf{R}^3_+ \cup \Omega)/G$ (see subsection 1) of a Kleinian group G constructed by Klein combination of groups G_1 and G_2 is a disk of the manifolds $M(G_1)$ and $M(G_2)$.

c) *Excision of knots and links.* Consider the three-dimensional sphere $S^3 = \mathbf{R}^3 \cup \{\infty\}$. Let L be a knot (a simple loop) or a link (a union of disjoint simple loops) in S^3. Removing L from S^3, we have an open three-dimensional manifold $M = S^3 \backslash L$. A large number of different open manifolds can be obtained in this way. An answer to the question of which of these manifolds admit a hyperbolic metric and, consequently, can be uniformized by Kleinian groups acting in $\mathbf{R}^3_+$, was formulated by Riley [211] as a conjecture: *the complement $S^3 \backslash L$ of a knot L admits a hyperbolic metric if and only if L is neither a torus knot nor a satellite knot.*

Here a torus knot is understood to be a knot that can lie on the ordinary torus T^2 in $\mathbf{R}^3$, and a satellite knot is understood to be a knot obtained by a nontrivial imbedding of the circle S^1 in a small toroidal neighborhood of some nontrivial knot K, i.e., an imbedding nonisotopic to K and with image not contained in a ball contained in $S^3 \backslash K$.

This conjecture was proved as a corollary of a uniformization theorem announced by Thurston [303].

Manifolds of this form are given in examples in §5 of Chapter III.

d) *Dehn surgery.* Let L be a knot or link in S^3. We excise a tubular neighborhood of L (whose boundary is obviously a torus) and glue it back with the help of some homeomorphism of the torus onto itself. Such a procedure has come to be called *Dehn surgery.* Since a torus admits sufficiently many homeomorphisms onto itself, this procedure also enables us to get a large number of nonhomeomorphic manifolds. With the help of Dehn surgery Thurston [234] was able to construct hyperbolic manifolds with very diverse properties, for example, the property of not being sufficiently large.

e) *Heegaard splittings.* Let M be a closed three-dimensional manifold. A *Heegaard splitting* of M is defined to be a pair (T_1, T_2), where T_i is a handlebody ($i = 1, 2$), and M is the union $T_1 \cup T_2$, with $T_1 \cap T_2 = \partial T_1 = \partial T_2$. (Here and below, a handlebody is understood to be an orientable manifold with boundary that is homeomorphic to a disk sum of solid tori.)

It is known [100] that each closed three-dimensional manifold has a Heegaard splitting, i.e., can be glued together from two homeomorphic handlebodies. Since the set of gluing homeomorphisms in the general case has not lent itself to description, it is impossible, as a rule, to say anything about a manifold M glued together in such a way. Certain simple cases are exceptions, for example, splittings of solid tori. Namely, suppose that the closed three-dimensional manifold M has a Heegaard splitting of genus 1. If M is orientable, then it can be shown to be either a *lens space* $L(p, q)$, where p and q are relatively prime positive integers (see Example 72) or homeomorphic to $S^2 \times S^1$.

§8. Uniformization of multidimensional manifolds

Unlike the case of Riemann surfaces, very little is known about uniformization of n-dimensional manifolds ($n > 2$) by Kleinian groups. Here we give some known results in this direction. The concept of a conformal manifold is a natural generalization of the concept of a Riemann surface to the multidimensional case. We first define the general concept of an (X, G)-manifold.

A pair (X, G) will be called a *model pair* if X is a real analytic manifold and G is a group of real analytic diffeomorphisms of X. Suppose now that M is a topological manifold. An (X, G)-*atlas* on M is defined to be a family $\{u_\alpha, \varphi_\alpha\}_{\alpha \in A}$, where $\{u_\alpha\}_{\alpha\in A}$ is an open covering of M and $\varphi_\alpha: u_\alpha \to D_\alpha$ is a homeomorphism of u_α onto an open subset $D_\alpha \subset X$, if for all $\alpha, \beta \in A$ with $u_\alpha \cap u_\beta \neq \varnothing$ the mapping $\varphi_{\alpha\beta} = \varphi_\alpha \circ \varphi_\beta^{-1}: D_\alpha \cap D_\beta \to D_\alpha \cap D_\beta$ is the restriction of an element of the group G. Any maximal (X, G)-atlas on M is called on (X, G)-*structure* on M, and manifolds on which some (X, G)-atlas can be defined are called (X, G)-*manifolds.* The manifold M together with some maximal atlas K will be denoted by (M, K). Let (M_1, K_1) and (M_2, K_2) be two (X, G)-manifolds. A local homeomorphism $f: M_1 \to M_2$ is called an (X, G)-*mapping* if the (X, G)-structure on M_1 obtained by lifting the structure K_2 by the mapping f is equivalent to K_1, i.e., the union of K_1 and $f^{-1}(K_2)$ is an (X, G)-atlas on M_1.

We introduce the concept of a development of an (X, G)-structure on a manifold M. Let (M, K) be an (X, G)-manifold. We consider the universal covering $p: \tilde{M} \to M$ and define on $\tilde{M}$ an (X, G)-structure $\tilde{K}$ as the lifting by the mapping p. By using properties of analytic diffeomorphisms it can be shown that there exist a local diffeomorphism $d: \tilde{M} \to X$ and a homomorphism $d^*: \pi_1(M) \to G$ ($\pi_1(M)$ is regarded as a group of covering homeomorphisms of the covering $p: \tilde{M} \to M$) such that $d(\gamma(x)) = d^*(\gamma)(d(x))$ for all $x \in \tilde{M}$ and $\gamma \in \pi_1(M)$.

The mapping d is an (X, G)-mapping and is called a *development* of the (X, G)-structure on M, and d^* is called the *holonomy homomorphism* and $\Gamma = d^*(\pi_1(M))$ the *holonomy group* of the structure K. A development of a fixed (X, G)-structure on M is determined to within composition with elements of G, and the holonomy group is determined to within conjugation in G.

The manifold M is said to be *conformal* if M is modeled by the pair $(S^n, \mathcal{M}(n))$, where $S^n = \overline{R}^n$ is the n-dimensional Möbius space, and $\mathcal{M}(n)$ is the group of Möbius automorphisms of S^n. The corresponding (X, G)-structure is said to be conformal, and the (X, G)-mappings are said to be conformal mappings.

REMARK. For $n = 2$ our definition of a conformal structure coincides with the definition of a projective structure on a surface, since in this case $\mathcal{M}(2)$ can be identified in a natural way with the group $PSL(2, \mathbf{C})$.

The class of conformal manifolds is sufficiently large. For example, it contains all the Riemannian conformally Euclidean manifolds (a Riemannian manifold is said to be *conformally Euclidean* if each point has a neighborhood which maps homeomorphically and conformally onto an open subset of a Euclidean space); in particular, all the Riemannian manifolds of constant sectional curvature. The class of Kleinian manifolds is the most important class of conformal manifolds. The former are defined to be the manifolds of the form Δ/Γ, where Γ is a torsion-free Kleinian group in $\mathcal{M}(n)$, and Δ is an invariant component of the set of discontinuity of Γ. Since Γ acts on Δ by conformal automorphisms, the conformal structure in Δ is projected onto Δ/Γ.

The description of the class of conformal manifolds represents a very difficult problem. For example, only recently Goldman [271] constructed examples of closed three-dimensional manifolds which do not have conformal structures.

The articles [283], [143], and [144] of Kuiper were the first to investigate conformal manifolds (Riemannian conformally Euclidean manifolds). These results are based on a study of the properties of a development and can be formulated as the following theorems.

THEOREM 1. *Let M be an n-dimensional ($n \geq 3$) closed conformal manifold with finite fundamental group. Then a development of any conformal structure K on M is a homeomorphism onto S^n, and (M, K) is conformally equivalent to S^n/Γ, where Γ is a finite group of Möbius automorphisms of S^n acting freely on S^n.*

THEOREM 2. *Let M be an n-dimensional ($n \geq 3$) closed conformal manifold with an infinite abelian fundamental group. If K is any conformal structure on M, then a development of it is a homeomorphism onto one of the following regions:* 1) $D = S^n \backslash \{x\}$; 2) $S^n \backslash \{x, y\}$. *In the first case the holonomy group Γ is conjugate in $\mathcal{M}(n)$ to a group of Euclidean motions, and (M, K) is conformally equivalent to $D/\Gamma = T^n$, where T^n is the n-dimensional Euclidean torus. In the second case the holonomy group is conjugate in $\mathcal{M}(n)$ to a group of similarities, and (M, K) is conformally equivalent to $D/\Gamma = S^{n-1} \times S^1$ with the structure lowered from D.*

THEOREM 3. *Let M be an n-dimensional ($n \geq 3$) closed manifold. If M is modeled by the pair $(R^n, \operatorname{Sim} R^n)$, where $\operatorname{Sim} R^n$ is the group of similarities of R^n, then a development of this structure is a homeomorphism onto one of the following regions:* 1) $D = S^n \backslash \{x\}$; 2) $D = S^n \backslash \{x, y\}$. *In the first case the holonomy group is a finite extension of a free abelian group of rank n that is conjugate in $\mathcal{M}(n)$ to a group of Euclidean motions, and M is a Bieberbach manifold, i.e., has a finite-sheeted covering by the torus T^n. In the second case the holonomy group is a finite extension of an infinite cyclic* *group and is conjugate in $\mathcal{M}(n)$ to a group of similarities which do not preserve the Euclidean metric, while M is an almost Hopf manifold, i.e., has a finite-sheeted covering by the Hopf manifold $S^{n-1} \times S^1$.*

In 1983 Goldman [271] obtained a generalization of these results. In particular, he showed that if the fundamental group of a closed conformal manifold is either solvable or nilpotent, then it is a finite extension of an abelian group, and, consequently, Kuiper's theorems give properties of the developments and a description of the holonomy groups.

A subsequent study of the properties of developments of conformal structures on three-dimensional manifolds was carried out by Gusevskiĭ and Kapovich. To formulate these results we give some definitions.

Let H^2 be the two-dimensional Lobachevsky space with the Riemannian metric $ds_H^2 = (dx_1^2 + dx_2^2)/x_2^2$ (H^2 is assumed to be realized as the upper half-plane $\{(x_1, x_2) \in R^2,\ x_2 > 0\}$), where R is the Euclidean line with the Riemannian metric dt^2. We consider the space $Z = H^2 \times R$ with the Riemannian metric $ds_Z^2 = ds_H^2 + dt^2$. Let $I(Z)$ be the group of isometries of Z. If a three-dimensional manifold M is modeled by the pair $(Z, I(Z))$ and M is closed, then M is a Seifert manifold. Moreover, since ds_Z^2 is conformally Euclidean, every $(Z, I(Z))$-manifold is equipped with a certain conformal structure given by means of ds_Z^2.

THEOREM 1. *Let M be a closed three-dimensional conformal manifold with infinite fundamental group and conformal structure K. Then the following two conditions are equivalent:*

a) $d(\tilde{M}) = D \neq S^3$.

b) *$d\colon \tilde{M} \to D$ is a covering.*

If, furthermorc, K is not equivalent to any $(Z, I(Z))$-structure on M, then each of the conditions a) *and* b) *is equivalent to*

c) *The holonomy group Γ is Kleinian, and $D = d(\tilde{M})$ is an invariant component of Γ.*

This implies that conformal structures with one of the conditions a) or b) but not c) can exist only on Seifert manifolds modeled by the pair $(Z, I(Z))$.

THEOREM 2. *Let M be a closed three-dimensional manifold, and let K be a conformal structure on M. If K is equivalent to some $(Z, I(Z))$-structure on M, then the following assertions are true:*

1) *The holonomy group Γ of this structure has an invariant circle $L(\Gamma)$ in S^3.*

2) $d(\tilde{M}) = S^3 \backslash L(\Gamma)$.

3) *Γ is either discrete or almost discrete (i.e., its restriction to $L(\Gamma)$ is discrete), and both cases are realized.*

4) *M has a finite-dimensional covering homeomorphic to $S_g \times S^1$, where S_g is a closed surface of genus $g \geq 2$ and S^1 is the circle.*

Let (M, K) be an n-dimensional $(n \geq 3)$ conformal manifold. We say that (M, K) is uniformized by a Kleinian group $\Gamma \subset \mathcal{M}(n)$ if Γ has an invariant component Δ of the set of discontinuity, and the manifold Δ/Γ with the natural conformal structure is conformally equivalent to (M, K). Kuiper's theorems show that *if M has a finite or almost abelian fundamental group or is modeled by the pair $(R^n, \operatorname{Sim} R^n)$, then any conformal structure on M is uniformized by the holonomy group of this structure.* In the general case there are no theorems of this type, and, as a rule, even if M has uniformizable structures, then M also has nonuniformizable structures. All this says that the uniformization problem has a negative solution in general for conformal structures on manifolds of dimension $n \geq 3$ in the classical formulation, and the theory of uniformization in the multidimensional case must apparently be similar to the theory of projective structures on a surface (see [277] and [280]).

CHAPTER III

Examples

This chapter has a dual purpose. It contains examples illustrating the general theory expounded in the first two chapters; on the other hand, most of these examples are of independent interest. They give solutions (not always positive) to the problems stated here, and many of them form the bases of methods used in the theory of Kleinian groups and the theory of uniformization.

The examples split naturally into several separate groups (often interconnected); they are divided into sections according to this principle. We continue the numbering of examples, the first fourteen of which were presented in the previous chapters.

§1. General properties of Kleinian groups (fundamental sets, sets of discontinuity)

EXAMPLE 15 (see, for example, Apanasov [31]). *A discrete but not discontinuous group in* $\mathbf{R}^n$, $1 \leq n \leq 4$. *The modular group.* Suppose that the group $\Gamma \subset \mathcal{M}_n$ is generated by the following mappings:

$$\begin{gathered} g_1(x) = x + e_1, \ldots, g_{n-1}(x) = x + e_{n-1}, \\ g_n(x) = |x|^{-2}(-x_1, x_2, \ldots, x_n), \qquad g_k(x) = U_k \cdot x, \\ k = n+1, \ldots, N, \end{gathered} \tag{44}$$

where $e_1, \ldots, e_{n-1}$ are the unit coordinate vectors in $\mathbf{R}^{n-1}$, and the U_k are all the orthogonal $n \times n$ matrices with elements $u^k_{ij} = \pm\delta_{ij}$ (δ_{ij} is the Kronecker symbol) such that $\det U_k = 1$.

It can be shown [31] that Γ acts discontinuously in $\mathbf{R}^n_+$. This follows from the fact that the polyhedron

$$\begin{aligned} F(\Gamma) = \{x \in \mathbf{R}^n_+ \colon |x| > 1;\ |x_{n-1}| < \tfrac{1}{2};\ 0 < x_i < \tfrac{1}{2}, \\ i = 1, \ldots, n-2\}, \end{aligned} \tag{45}$$

whose faces are identified pairwise by the generators (44) of Γ, satisfies the conditions of Poincaré's theorem. The hyperbolic polyhedron (45) is not compact, but for $2 \leq n \leq 5$ it has finite volume, since for $2 \leq n \leq 4$ it has a single parabolic vertex at ∞, while for $n = 5$ there are two additional parabolic vertices lying on the boundary plane $\mathbf{R}^{n-1}$. This implies that Γ is a discrete subgroup of the Möbius group $\mathcal{M}_n$ (this can also be obtained from the fact that Γ has a subgroup of finite index to whose elements there correspond matrices with integer entries in the Lorentz group representation (see §2 of Chapter I, and also [28])).

The restriction of the action of Γ to its invariant subspace $\mathbf{R}^{n-1}$ is again a discrete group $G \subset \mathcal{M}_{n-1}$, but it does not act discontinuously there for $2 \leq n \leq 5$. This follows, in particular, from the fact that the limit set $\Lambda(G)$ of G coincides with the limit set $\Lambda(\Gamma)$ of Γ (which has a fundamental polyhedron (45) of finite volume and, consequently, is a Fuchsian group of the first kind in $\mathbf{R}^n$), which, in turn, coincides with the boundary plane $\mathbf{R}^{n-1}$.

For $n = 2$ the group Γ coincides with the usual modular group of all linear fractional mappings of the form $g(z) = (az + b)/(cz + d)$, where a, b, c, and d are real integers such that $ad - bc = 1$.

For $n = 3$ the restriction of Γ to the plane $\mathbf{R}^2 = \mathbf{C}$, i.e., the group G, coincides with the well-known Picard group of all mappings of the form $g(z) = (az+b)/(cz+d)$, where a, b, c, and d are complex integers such that $ad-bc = 1$.

For $n \geq 6$ the restriction of Γ to $\mathbf{R}^{n-1}$, i.e., the group G, is now a Kleinian group in $\mathbf{R}^{n-1}$ (whose set of discontinuity does not have any invariant components).

EXAMPLE 16 (GREENBERG [86]). *A nonabelian purely loxodromic group in a half-space that is not discrete.* Suppose that a group $G \subset \mathcal{M}_n$ acts in the ball B^n or the half-space $\mathbf{R}^n_+$. An axis of a loxodromic (hyperbolic) element g of it is defined to be a non-Euclidean line $l \subset B^n$ (or $\mathbf{R}^n_+$) with $g(l) = l$. It is known [77], [29] that the following conditions are sufficient for such groups to be discrete: a) G is a nonabelian group, and b) G consists solely of hyperbolic elements. Moreover, the condition that G be nonabelian is equivalent to the condition that G have more than one axis.

It would be natural to expect that any purely loxodromic group in a ball (half-space) with more than one axis is discrete. The following example shows that this is not so for $n \geq 3$.

We remark that the restriction of a mapping $g \in \mathcal{M}_3, g(\mathbf{R}^3_+) = \mathbf{R}^3_+$ to the plane $\overline{\mathbf{C}} = \partial\mathbf{R}^3_+$ is a linear fractional mapping $\gamma(z) = (az + b)/(cz + d)$. This mapping $\gamma(z)$ and, hence, also $g(x)$ are loxodromic only when the trace $\operatorname{tr}(\gamma) = a + d$ is not real.

Denote by $\{U_m\}_{m=1,2,\dots}$ a base of open sets in $\mathbf{R}^6$ and consider a sequence $\{P_m\}$ of points $P_m = (a'_m, a''_m, b'_m, b''_m, c'_m, c''_m)$ in $\mathbf{R}^6$ such that $P_m \in U_m$ for all m, and the components of all the P_i, $1 \leq i \leq m$, are independent transcendentals (over the rational field). Let

$$a_m = a'_m + ia''_m,\ b_m = b'_m + ib''_m,\ c_m = c'_m + c''_m,\quad d_m = 1 + b_m c_m / a_m \tag{46}$$

and consider the group $G = \langle g_1, \dots, g_m, \dots \rangle$, generated by the mappings $g_m(z) = (a_m z + b_m)/(c_m z + d_m)$ with coefficients in (46). It is clear from the construction of G that it is dense in $\mathcal{M}_2$ and, consequently, is nondiscrete. We show that G is purely loxodromic. To do this it suffices to show that for any $g \in G$ either $\operatorname{tr}(g) \notin \mathbf{R}$ or $g = I$ $(g(z) \equiv z)$. Indeed, g is a word $w(g_{k_1}, \dots, g_{k_n})$ in the generators of G, and

$$\operatorname{tr}(g) = a + d = \mathcal{R}(a_{k_1}, b_{k_1}, c_{k_1}, \dots, a_{k_n}, b_{k_n}, c_{k_n}), \tag{47}$$

where $\mathcal{R} = U + iV$, with U and V real rational functions. If $V = 0$ for g, then $V \equiv 0$, because the components of the P_i are independent transcendentals. But then the analyticity of $\mathcal{R}$ gives us that $\mathcal{R}$, which is defined by (47), is identically equal to a real constant ρ. This means that $\operatorname{tr}(\gamma) = \rho$ for any $\gamma_1, \dots, \gamma_n \in \mathcal{M}_2$, where $\gamma = w(\gamma_1, \dots, \gamma_n)$. Setting $\gamma_1 = \gamma_2 = \cdots = \gamma = I$, we get that $\rho = 2$. If now $\gamma_1, \dots, \gamma_n$ are free generators of a free hyperbolic group, then the equality $\operatorname{tr}(\gamma) = 2$ implies that $\gamma = I$, i.e., that $w(\gamma_1, \dots, \gamma_n) = I$. From this it follows that the word w can be reduced freely to **1** and, consequently, $g = w(g_{k_1}, \dots, g_{k_n}) = I$.

It has thus been shown that our group G consists of loxodromic elements. Extending them to the half-space $\mathbf{R}^3_+$, we obtain a purely loxodromic group $\hat{G} \subset \mathcal{M}_3, \hat{G}(\mathbf{R}^3_+) = \mathbf{R}^3_+$, that has a countable set of distinct axes (and hence is nonabelian) but is not discrete.

The next two examples are elementary, but the construction used in them is very useful.

EXAMPLE 17 (cf. Ford [78]). *Kleinian groups whose isometric fundamental regions have countable sets of components.* Take a countable set of spheres $I_1, I'_1, I_2, I'_2, \dots$ lying in the finite part of the plane $\mathbf{R}^n$ $(n \geq 2)$ and situated outside one another except possibly for tangency from the outside, and with each pair I_k, I'_k of equal radius. Consider now Möbius mappings g_k having, together with their inverses g_k^{-1}, the isometric spheres I_k and I'_k, respectively. Then the free group $G = \langle g_1, g_2, \dots \rangle$ (Klein combination) generated by these mappings is discontinuous in $\mathbf{R}^n$ and has as fundamental polyhedron P_G the exterior of all the spheres I_k, I'_k. This polyhedron splits into a countable number of parts when the spheres are suitably located.

Another example of a Kleinian group G with a similar property of the isometric fundamental polyhedron P_G is obtained when we consider in $\mathbf{R}^n$ a sequence $\{G_m\}$ of Fuchsian groups of the first kind that leave invariant balls $\{B_m\}$ that are situated sufficiently far from one another and when the group G is taken to be the free product of $\{G_m\}$, i.e., a group obtained from the G_m $(m = 1, 2, \dots)$ by Klein combination (see §3 in Chapter I). The group

G thus obtained has a single invariant component, and a countable set of nonconjugate noninvariant components.

EXAMPLE 18 (FORD [78]). *A Kleinian group such that the boundary of the isometric fundamental region consists of limit points.* Let C be a closed Jordan curve in $\overline{\mathbf{C}}$. Take a filling of the interior of C by countably many disks $\{U\}$ such that for each disk $U_k \in \{U\}$ there exists a disk U_k' of the same radius, and such that these disks accumulate at all the points of C and have union that is dense in $\operatorname{int} C$. For each pair U_k, U_k' we construct a linear fractional transformation g_k such that g_k and its inverse g_k^{-1} have the boundaries of U_k and U_k' as isometric circles, respectively. Consider the group $G = \{g_1, g_2, \dots\}$ generated by these transformations. Then $C \subset \Lambda(G)$, and the exterior of C is the isometric fundamental region P_G of G.

It is clear that an analogous example can be constructed also for any space $\overline{\mathbf{R}}^n$ $(n > 2)$ by taking a closed Jordan surface in place of C.

The exterior of C gives an example of a (simply connected) component of a Kleinian group G with trivial stabilizer in G. Such components are called *atoms*.

EXAMPLE 19 (APANASOV). *A doubly connected Kleinian group that is a $\mathbf{Z}_2$-extension of a quasi-Fuchsian group and has isometric fundamental polygon consisting of four components.* It is well known that the isometric fundamental polygon of a Kleinian group Γ with a circle S as its limit set has exactly two components, which lie inside and outside S, respectively. In this connection the question was posed (see [147, p. 136]) as to whether this is so for all Kleinian groups whose sets of discontinuity are divided into two components by a limit quasicircle. Jørgensen [108] proved that this is not true for all quasi-Fuchsian groups, but no explicit examples were constructed.

We construct a doubly generated Kleinian group G with the following properties: 1) $\Lambda(G)$ is a quasicircle.

2) The isometric fundamental polygon $P(G)$ in $\Omega(G)$ splits into four components P_∞, P_1, P_2, P_3.

3) G contains a quasi-Fuchsian subgroup $G_0 \subset G$ of index 2 such that $P(G_0)$ has three components.

4) G is obtained from a group Γ with $\Lambda(\Gamma)$ a circle by quasiconformal deformation.

5) The space $\Omega(G)/G$ of orbits is a Riemann surface homeomorphic to a torus with two points deleted.

The two loxodromic mappings

$$g_1(z) = 1/(z+1), \qquad g_2(z) = (2z+2+5i)/(iz-2+i) \tag{48}$$

with traces $\operatorname{tr} g_1$ and $\operatorname{tr} g_2$ of modulus 1 are taken as the generators of the group G. The latter is the free product of the cyclic subgroups $\{g_1\}$ and $\{g_2\}$, and its isometric fundamental polygon $P(G)$ (Figure 12) has four connected components. Of the four points of tangency of the isometric circles of $\{g_1\}$ an

$\{g_2\}$ only the extreme ones are limit points. This is easily seen by considering the image $g_1(P_1 \cup P_2 \cup P_3)$ and $g_1^{-1}(P_1 \cup P_2 \cup P_3)$.

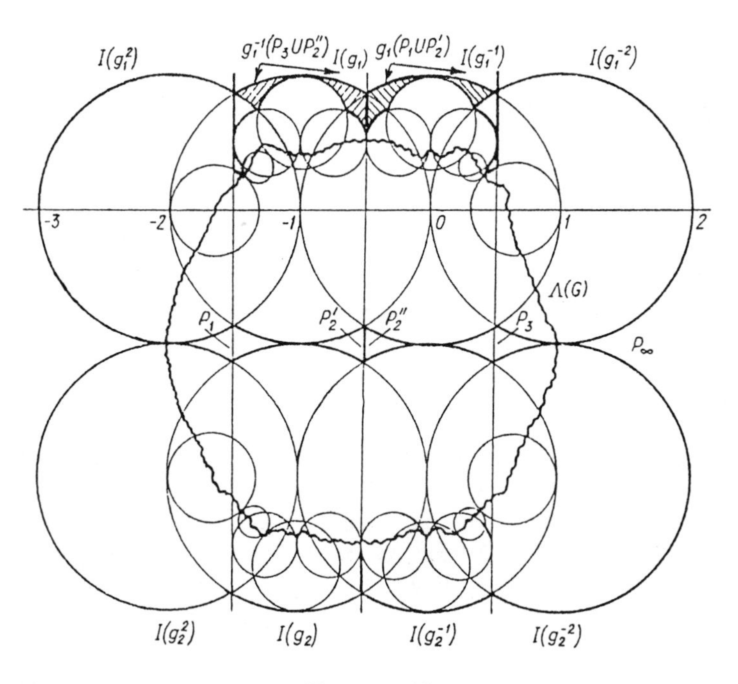

FIGURE 12

As the group Γ to be deformed ($G = f\Gamma f^{-1}$) we take the free product of the cyclic groups $\{\gamma_1\}$ and $\{\gamma_2\}$, where

$$\gamma_1(z) = \frac{-(11+5i)z + 9\frac{2}{5} - 5i}{12z + 1 + 5i},$$
$$\gamma_2(z) = \frac{-(1+25i)z + 59\frac{2}{5} - 25i}{12z + 11 + 25i}. \tag{49}$$

Figure 13 shows the fundamental polygon $P(\Gamma) = \tilde{P}_\infty \cup (P_0' \cup P_0'')$ (which is not compact, has finite hyperbolic volume, and splits into two components). To construct a quasiconformal mapping $F\colon \overline{\mathbf{C}} \to \overline{\mathbf{C}}$ with $G = f\Gamma f^{-1}$ we take a somewhat varied fundamental region of Γ, namely,

$$F(\Gamma) = \tilde{P}_\infty \cup \gamma_1(P_0') \cup \gamma_1^{-1}(P_0''). \tag{50}$$

Correspondingly, we take the fundamental region of G to be

$$F(G) = P_\infty \cup g_1(P_1 \cup P_2') \cup g_1^{-1}(P_3 \cup P_2''). \tag{51}$$

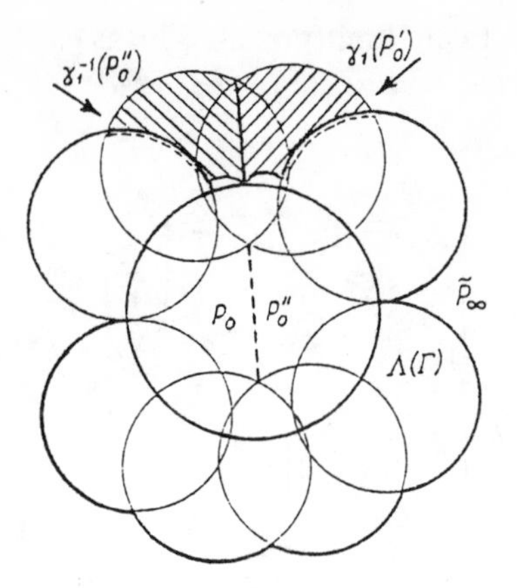

FIGURE 13

If now $f_1: F(\Gamma) \to F(G)$ is a quasiconformal mapping carrying the sides of $F(\Gamma)$ into the corresponding sides of $F(G)$, then the desired mapping $f: \overline{\mathbf{C}} \to \overline{\mathbf{C}}$ is obtained from f_1 by extending "by periodicity" (taking into account the removability of the circle $\Lambda(\Gamma)$ for f). Since $G = f\Gamma f^{-1}$ and $\Lambda(\Gamma) = \{z \in \mathbf{C}: |z + 1/2 + 5/4i| = 5/4\}$, it follows that $\Lambda(G) = f(\Lambda(\Gamma))$ is a quasicircle and $\Omega(G) = \Omega_0 \cup \Omega_1$.

It is not hard to show that there exist a quasi-Fuchsian subgroup $G_1 \subset G$ and an element $g \in G$ such that $G = G_1 \cup gG_1$, and the isometric polygon P_{G_1} of this subgroup has three components: $P_{G_1} = P_\infty(G_1) \cup P_1(G_1) \cup P_2(G_1)$.

EXAMPLE 20 (ACCOLA [9]). *A Kleinian group with two invariant components and with atoms.* Let $K = \{z \in \mathbf{C}: 1 < |z| < 2\}$, and let C_1 and C_2 be two infinite disjoint spirals in K converging to $\{|z| = 1\}$ and $\{|z| = 2\}$. They divide K into two simply connected regions R_1 and R_2 with the common boundary $\partial R_1 = \partial R_2 = C_1 \cup C_2 \cup \{|z| = 1\} \cup \{|z| = 2\}$.

Let $\{S_{i,k},\ k \in \mathbf{Z}\}$, $i = 1, 2$, be two families of circles with the following properties:

a) $S_{i,k-1}$ is tangent to $S_{i,k}$ at a point of the spiral C_i, $i = 1, 2$.

b) $S_{1,k} \cap S_{2,j} = \varnothing$ for all k, $j \in \mathbf{Z}$.

c) The exterior of all the circles $\{S_{i,k}\}$ splits into four regions $\{|z| < 1\}$, $\{|z| > 2\}$, F_1, and F_2; here the $F_i = R_i \cap \operatorname{ext} S_{j,k}$, $i = 1, 2$, are subregions of R_i.

Let the group G' be generated by all the inversions $J_{i,k}$ with respect to the circles $S_{i,k}$, $i = 1, 2, k \in \mathbf{Z}$, and let $G \subset \mathcal{M}_2$ be the subgroup of G' of index 2 consisting of all the linear fractional mappings in G'. Then $F_1 \cup J_0(F_1)$ and $F_2 \cup J_0(F_2)$ are connected components of a fundamental region of G (J_0 is inversion with respect to some circle S_{i_0,k_0}) that lie in distinct components Ω_1 and Ω_2 of the set of discontinuity. The group G' is such that any

$p_i \in F_i$ $(i = 1,2)$ can be joined to any of its images $J_{j,k}(p_i)$, $i = 1,2$, $j = 1,2, k \in \mathbf{Z}$, by a path in $\Omega(G) = \Omega(G')$. Therefore, G has exactly two invariant components

$$\Omega_1 = \bigcup_{g\in G} g\{F_1 \cup J_0(F_1)\} \quad \text{and} \quad \Omega_2 = \bigcup_{g\in G} g\{F_2 \cup J_0(F_2)\}.$$

The remaining part of the set of discontinuity splits into atoms:

$$\Omega(G)\backslash(\Omega_1 \cup \Omega_2) = \bigcup_{k=3}^{\infty} \Omega_k, \tag{52}$$

where all the connected components Ω_k $(k \geq 3)$ of the set in (52) are simply connected, have trivial stabilizers in G, and are the images of the regions $\{|z| < 1\}$ and $\{|z| > 2\}$ under mappings in G. We note once more that, by Ahlfors' theorem, such pathologies do not occur for finitely generated groups (see §5 in Chapter I).

EXAMPLE 21 (APANASOV [36]). *Finitely generated Kleinian groups in space that have exactly two invariant components such that the simple connectness property is lost.* As a first example of such a group (for simplicity we set $n = 3$) we can take a group $G_1 \subset \mathcal{M}_3$ lying on the boundary $\partial T(G)$ of the Teichmüller space $T(G)$ of the Fuchsian group G of the first kind in Example 35. It is a limit point for the sequence of quasiconformal deformations $G(r, \tau, t)$ of a Fuchsian group Γ, where the parameters r and τ are constant and equal to 5/6, and the parameter t tends to $t_1 = 5(1 + \sqrt{5/6})$. This group G_1 has a system of generators $g_i, g_i^2 = 1, 1 \leq i \leq 14$, whose isometric spheres I_j are such that I_{13} and I_{14} are tangent to the spheres I_9, I_{10}, I_{11}, and I_{12} outside the ball $B_0, G(B_0) = B_0$, while the remaining spheres are the same as for G. As in Example 35, it can be shown that the set $\Omega(G_1) = \Omega(G(5/6, 5/6, t_1))$ of discontinuity is the union of its two invariant connected components Ω_0 and Ω_1. The bounded component Ω_0 is the quasiconformal image of a ball and is thus simply connected. This cannot be said of the component Ω_1; its fundamental group $\pi_1(\Omega_1)$ is an infinitely generated free group.

A second Kleinian group $G_2 \subset \mathcal{M}_3$ with the analogous property is also on the boundary $\partial T(G)$ of the Teichmüller space of the Fuchsian group G in Example 35, and is the limit of the groups $G(r, \tau, t)$ when the parameter $r = 5/6$ there is constant, the parameter τ tends to

$$\tau_0 = 5(1 + \sqrt{5/6})/(11 + 12\sqrt{5/6}),$$

and t tends to $t_1 = 5(1 + \sqrt{5/6})$. As in the case of the groups G and G_1, it can be shown that the set $\Omega_2 = \Omega(G(5/6, \tau_0, t_1))$ of discontinuity is the union of its two invariant components Ω_0 and Ω_1. But unlike in the previous example, neither of them is now simply connected: $\pi_1(\Omega_0)$ and $\pi_1(\Omega_1)$ are infinitely generated free groups. The regions Ω_1 and Ω_2 can be characterized as bodies of infinite genus that are interlaced.

We remark that for finitely generated Kleinian groups on the plane the existence of two invariant components implies that they are simply connected, that there are no other components, and that the group itself is quasi-Fuchsian (see §5 in Chapter I, and also [9]).

EXAMPLE 22 (TETENOV [232], [233]). *The set of discontinuity of a spatial Kleinian group can consist of any number of invariant components, even simply connected ones.* This example shows how strongly the spatial case differs from the case of Kleinian groups on the plane. There, as is well known (see §1 in Chapter I), the set of discontinuity of a group has one or two or infinitely many connected components. Moreover, among them no more than two can be invariant, and in the case of two invariant components both are simply connected (cf. the preceding example).

The construction of our example reduces to the construction of an infinite family Σ of spheres in space that satisfy the following conditions:

a) $\operatorname{int}\sigma_i \cap \operatorname{int}\sigma_j = \varnothing$ for all distinct $\sigma_i, \sigma_j \in \Sigma$.

b) The set $\bigcap \operatorname{ext}\sigma_j$ splits into k connected components, namely, the spherical polyhedra $P_1, \dots, P_k$.

c) For any $\sigma_j \in \Sigma$ and for any $m = 1, \dots, k$ the set $\sigma_j \cap P_m$ has a nonempty interior in σ_j.

d) The polyhedra P_m are homeomorphic to a ball, and $\sigma_j \cap \overline{P}_m$ is a simply connected region on the sphere $\sigma_j \in \Sigma$, $m = 1, \dots, k$.

Let $G = G(\Sigma)$ be the group generated by compositions of an even number of inversions with respect to spheres in the family Σ. It follows from condition a) that G is discontinuous. Let Ω_m be the set obtained from P_m by reflections with respect to all the spheres $\sigma_j \in \Sigma$ (each of them contains some face of P_m). It follows from conditions a) and c) that each of the sets $\Omega_m, m = 1, \dots, k$, is an invariant component of G. The union of all the Ω_m gives the set of discontinuity of G. It follows from condition d) that each component Ω_m is homeomorphic to a ball. Thus, the group $G(\Sigma)$, which satisfies conditions a)–d), is the desired group.

For simplicity we confine ourselves to the case when $k = 3$ and $n = 3$. We first construct a family Σ satisfying conditions a)–c), and we then show how to satisfy condition d). Let us begin by choosing two closed disjoint surfaces S_0^1 and S_0^2 in space, and "approximating" each of them by a family of spheres such that 1) the union of the balls bounded by these spheres contains the given surface and lies in some neighborhood of it, and 2) the points of tangency of the spheres in this family lie on the given surface.

We thereby obtain a family Σ_0 of spheres satisfying conditions a) and b). The set $\bigcap_{\sigma_j \in \Sigma_0} \operatorname{ext}\sigma_j$ splits into three connected components: the spherical polyhedra P_0^0, P_0^1, and P_0^2. Each of the spheres $\sigma_j \in \Sigma_0$ borders on only two of the polyhedra: the unbounded one P_0^0 and one of the bounded ones. For c) to hold it is necessary to "join" each such sphere with that one of the polyhedra P_0^1 or P_0^2 which was not previously its neighbor. To do this we remove from the family Σ_0 one sphere bordering on P_0^1 and one bordering on P_0^2. This gives a family $\tilde{\Sigma}_0$. The set $\bigcap_{\sigma_j\in\tilde{\Sigma}_0}\operatorname{ext}\sigma_i$ is connected, and holes are formed in the boundaries of P_0^1 and P_0^2. On each of the spheres $\sigma_j\in\tilde{\Sigma}_0$ we take a part δ_j bounded by a curve γ_j such that $\delta_j\subset\sigma_j\cap P_0^0$. We now construct two disjoint surfaces S_1^1 and S_1^2 lying in P_0^0 and such that each surface S_1^r $(r=1,2)$ is a branched tube joining the hole in the boundary of P_0^r to the loops γ_j on the spheres σ_j that bound the polyhedron P_0^{3-r}. Then the two indicated surfaces S_1^1 and S_1^2 divide the set $\bigcap_{\sigma_j\in\tilde{\Sigma}_0}\operatorname{ext}\sigma_j$ into three components, and each of the spheres $\sigma_j\in\tilde{\Sigma}_0$ borders on all these components. Each of the surfaces S_1^1 and S_1^2 can be approximated by some family Σ_1 of spheres (see 1) and 2)). This family can be chosen so that for any $\sigma_i\neq\sigma_j,\sigma_i,\sigma_j\in\Sigma_1\cup\tilde{\Sigma}_0$, the intersection of σ_i and σ_j is empty, and

$$\bigcap_{\sigma_i\in\Sigma_1\cup\tilde{\Sigma}_0}\operatorname{ext}\sigma_i=P_0^0\cup P_0^1\cup P_0^2. \tag{53}$$

In other words, the family $\Sigma_1\cup\tilde{\Sigma}_0$ satisfies conditions a) and b). Property c) is satisfied only for the spheres $\sigma_i\in\tilde{\Sigma}_0$. But each of the spheres $\sigma_i\in\Sigma_1$ borders on only two polyhedra P_1^0 and P_1^r, where $r=1$ or 2. Removing one sphere of the family Σ_1 from each of the boundaries of P_1^1 and P_1^2, we obtain a new family $\tilde{\Sigma}_1$ of spheres. Repeating this process (53) infinitely many times, we arrive at a sequence of families Σ_m and $\tilde{\Sigma}_m$ of spheres such that

$$\bigcap\operatorname{ext}\sigma_k=P_m^0\cup P_m^1\cup P_m^2. \tag{54}$$

Here the intersection is over all the spheres $\sigma_k\in\bigcup_{n=0}^{m-1}\tilde{\Sigma}_n\cup\Sigma_m$.

Each of the families $\bigcup_{k=0}^{m-1}\tilde{\Sigma}_k\cup\Sigma_m$ satisfies conditions a) and b), and condition c) fails only for $\sigma_j\in\Sigma_m$. Let $\Sigma=\bigcup_{k=0}^{\infty}\tilde{\Sigma}_k$, and let $P^r=\bigcup_{m=0}^{\infty}\bigcap_{k=m}^{\infty}P_k^r$, for $r=0,1,2$. The polyhedra P^1 and P^2 are connected, since they are constructed as unions of increasing sequences of connected polyhedra P_k^1 and P_k^2. It remains for us only to get the connectedness of $P^0=\bigcap_{k=0}^{\infty}P_k^0$. This can be achieved as follows: At each of our steps (54), before constructing the surfaces S_m^1 and S_m^2, we choose in P_{m-1}^0 a connected neighborhood V_{m-1} of its boundary and require that S_m^1, S_m^2, and the spheres of the family Σ_m be contained in V_{m-1}. The neighborhoods V_m are chosen so that $V_k\supset V_{k+1},\bigcap_{k=0}^{\infty}V_k=\varnothing$, and the sets P_k^0/V_k are connected. This results in sequences

$$\begin{array}{llll}
V_0\supset V_1\supset & & \cdots\supset\cdots= & \varnothing\\
\cap & & & \cap\\
P_0^0\supset P_1^0\supset & & \cdots\supset\cdots= & P^0\\
\cup & & & \cup\\
P_0^0\backslash V_0\subset P_1^0\backslash V_1\subset & & \cdots\subset\cdots\subset & P^0
\end{array}$$

Then P^0, as the union of the connected sets P^0_m/V_m, is connected. Thus, a family Σ satisfying conditions a)–c) is obtained (see Figure 14).

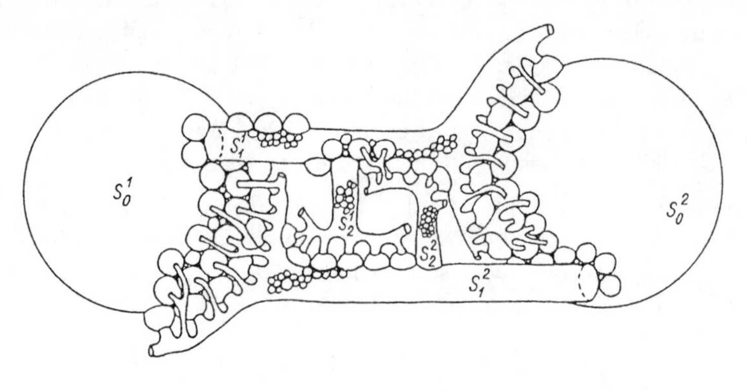

FIGURE 14

To satisfy condition d) we make some changes in our construction. First, take two spheres with a single point of tangency as S^1_0 and S^2_0. This enables us to construct a family Σ_0 such that the polyhedra P^0_0, P^1_0, and P^2_0 are homeomorphic to a ball. Second, in order that each of the surfaces S^k_m partition P^0_{m-1} into simply connected parts, we construct these surfaces S^k_m as follows: We join the hole in the boundary of P^k_{m-1} and the parts δ_i on the spheres $\sigma_i \in \tilde{\Sigma}_{m-1}$ bounding P^{3-k}_{m-1} by means of a branched curve ξ that lies in ∂P^0_{m-1} and is simply connected; "pushing" this curve onto the boundary of each of the sets $\sigma_i \cap \partial P^0_{m-1}$, we make sure that ξ does not partition any of these sets into parts. Then the surface S^k_m is glued to ∂P^0_{m-1} along the curve ξ.

Making the indicated changes at each step of the construction, we obtain a family Σ satisfying conditions a)–d), and thereby conclude the construction of the corresponding group G.

EXAMPLE 23 (ABIKOFF [4]). *A web group.* Consider the circles $C_1 = \{z\colon |z-1| = \sqrt{2}/2\}$, $C_2 = \{z\colon |z-i| = \sqrt{2}/2\}$, $C_3 = \{z\colon |z+i| = \sqrt{2}/2\}$ and $C_4 = \{z\colon |z+1| = \sqrt{2}/2\}$. Each is tangent to two of the others. In each disk $\operatorname{int} C_k$ we construct a Fuchsian group of the first kind with four parabolic generators. The fixed points of these generators lie at points of tangency and at points that are closest points and farthest points from the origin. By the first Maskit combination theorem we get that the group G', generated by these four Fuchsian groups, is Kleinian. Consider the free product of G' and the cyclic group generated by the mapping $\gamma(z) = (2+\sqrt{2})^2 z/2$; the group $G = \langle G', \gamma\rangle$ thus obtained is Kleinian. Moreover, any component of G is conjugate to one of the four disks of the original Fuchsian groups, i.e., G is web group(9). In Figure 15 the dashes denote circles bounding a fundamental polygon of G, and part of the limit set is represented by solid circles (the quadrangles between them are also filled by the images of their interiors).

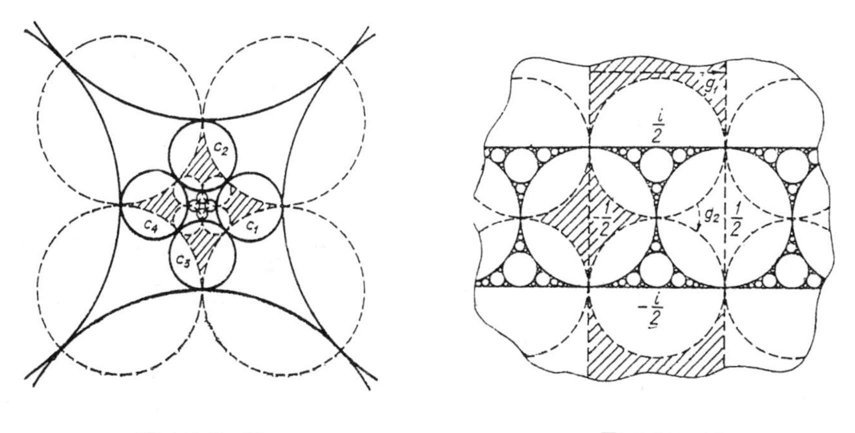

FIGURE 15 FIGURE 16

EXAMPLE 24. *A doubly connected web group.* As such a group we can take a Kleinian group that is the free product of two cyclic groups. One of them is generated by the translation $g_1(z) = z + 1$, and the second is generated by the parabolic transformation $g_2(z) = z/(2iz + 1)$, which leaves the point 0 fixed. The fact that this group is discontinuous follows from Klein's combination theorem. The set $\Omega(G)$ of discontinuity of this group splits into three classes of equivalent components. The half-plane $\{\operatorname{Im} z > 1/2\}$, the disk $\{|z - 1/2| = 1/2\}$, and the half-plane $\{\operatorname{Im} z < -1/2\}$ are representatives of these classes. Figure 16 gives the scheme for filling $\overline{\mathbf{C}}$ by their images and shows a fundamental polygon (it is shaded) of G.

EXAMPLE 25 (APANASOV) *Web groups in space.* We first give an example of an infinitely generated Kleinian group in $\overline{\mathbf{R}}^n$ (for simplicity let $n = 3$) with the same property of being a web group, i.e., such that the stabilizer of each of its components is a quasi-Fuchsian group. For this we consider the quasi-Fuchsian group $G_1 \subset \mathcal{M}_3$ in Example 42 (the "cube group") and the abelian group $G_1' \subset \mathcal{M}_3$ of rank 3 generated by translations by the unit coordinate vectors $e_1 = (1,0,0), e_2 = (0,1,0)$, and $e_3 = (0,0,1)$. The fundamental region of G_1 can be taken to be the set (71), which splits into two spherical polyhedra—a bounded one F_0 in the unit cube Q, and an unbounded one F_1 in the exterior of Q—and $F(G_1')$ can be take to be Q itself.

We define the group $G \subset \mathcal{M}_3$ to be the product of G_1 and G_1'; it is Kleinian. This follows already from the fact that there are no G-equivalent points in

(9) *Translation editor's note*: but not a free product.

the interior of F_0. Moreover, it can be shown (cf. [78], §35, Theorem 22) that F_0 is a fundamental region $F(G)$ of G. The set $\Omega(G)$ of discontinuity splits into countably many noninvariant components Ω_i $(i = 0, 1, \ldots)$, and

$$\Omega_0 = \bigcup_{g \in G_1} g(F_0), \quad \Omega_i = \gamma_i(\Omega_0), \qquad i \geq 1,\ \gamma_i \in G \backslash G_1. \tag{55}$$

It is clear that the stabilizer of any component Ω_i in (55) is the subgroup $\Gamma_i \subset G, \Gamma_i = \gamma_i G_1 \gamma_i^{-1}$, $i = 0, 1, 2, \ldots, \gamma_0 = I$, which is quasi-Fuchsian. Thus, G is a "web" group.

Let us pass to the needed example of a finitely generated web group in $\mathbf{R}^n$, $n \geq 3$. Without loss of generality we assume for simplicity that $n = 3$. The construction will use a construction analogous to that in Example 35. Suppose again that Q is the unit cube, and the S_i $(1 \leq i \leq 8)$ are spheres of radius $\sqrt{3}/3$ about its vertices. Take the spheres S_i $(9 \leq i \leq 14)$ to be the disjoint spheres of radius $1/2 - \sqrt{3}/6$ about the centers of the faces of Q. These spheres S_i $(1 \leq i \leq 14)$ intersect only at angles equal to $\pi/3$, and their interiors cover the surface of Q. It can be shown in a way analogous to Example 35 that the group G_1 generated by the involutions $g_i = O_i \cdot J_i$ with respect to the spheres S_i $(1 \leq i \leq 14)$ is quasi-Fuchsian and is, moreover, a quasiconformal deformation of some Fuchsian group in $\overline{\mathbf{R}}^3$ of the first kind. Furthermore, its fundamental region can be taken to be the exterior of the spheres S_i $(1 \leq i \leq 14)$, which has two components: $F_0 \subset Q$ and $F_1 \subset \overline{\mathbf{R}}^3/Q$. The set of discontinuity of G_1 consists of the two invariant components

$$\Omega_0 = \bigcup_{g \in G_1} g(F_0), \qquad \Omega(G_1) \backslash \Omega_0 = \bigcup_{g \in G_1} g(F_1). \tag{56}$$

Again we define the group G to be the product of G_1 and G_1', where G_1' is the abelian group of rank 3 constructed above. It can be shown in a way analogous to that in the preceding example that G is a web group.

The spherical polyhedron $F_0 \subset Q$ is a fundamental region of G, and the set $\Omega(G)$ of discontinuity is the union of its countably many noninvariant components,

$$\Omega(G) = \bigcup_{i=0}^{\infty} \Omega_i, \quad \Omega_0 = \bigcup_{g \in G_1} g(F_0), \quad \Omega_i = \gamma_i(\Omega_0), \quad \gamma_i \in G \backslash G_1, \tag{57}$$

where all the components Ω_i in (57) are equivalent to the component Ω_0 in (56). Therefore, the stabilizer of each component Ω_i is the quasi-Fuchsian subgroup $\Gamma_i \subset G$ obtained from G_1 by the conjugation $\Gamma_i = \gamma_i G_1 \gamma_i^{-1}$, $i \geq 1, \Gamma_0 = G_1$.

In both the first and the second example the web groups G have nonempty residual limit sets $\Lambda_0(G)$. In particular, they contain the orbit $\{g(\infty) \colon g \in G\}$ of the point at infinity.

EXAMPLE 26 (MASKIT [181]). *The conditions of the Maskit combination theorem cannot be weakened.* Let G_1 be the cyclic group generated by the transformation $g_1(z) = 3z$. Then $F_1 = \{z\colon 1 \leq |z| < 3\}$ is a fundamental set of it. Further, let G_2 be the group generated by the transformation $g_2(z) = (2z+5)/(z-2)$. As its fundamental set we take

$$F_2 = \{z\colon |z-2| \geq 1, |z+2| > 1, z \neq 1\} \cup \{z = -3\}. \tag{58}$$

The intersection F of the annulus F_1 with the region F_2 in (58) is a partial fundamental set for the group $G = \langle G_1, G_2 \rangle$. It can be shown that the real axis $\mathbf{R}$ is invariant under G, $\mathbf{R} \cap F = \varnothing$, and $\Omega(G) \cap \mathbf{R} \neq \varnothing$. This implies that F is not a full fundamental set for G. Therefore, G is not combinable from G_1 and G_2, which have fundamental regions F_1 and F_2.

EXAMPLE 27 (MASKIT [185], GUSEVSKIĬ [92]). *A finitely generated Kleinian group isomorphic to the fundamental group of a closed nonorientable surface.* Let $p \geq 1$, and let $4p+2$ circles of equal radius with centers on the positive real semiaxis be arranged so that the first is orthogonal to the circle $|z| = 1$, while the last is orthogonal to the circle $|z| = 2$ (all the remaining angles between the circles are equal). Let P be a polygon symmetric with respect to the imaginary axis which is bounded by arcs of these $4p+2$ circles, by the arcs symmetric to them with respect to the imaginary axis, and by arcs of the circles $|z| = 1$ and $|z| = 2$ (Figure 17).

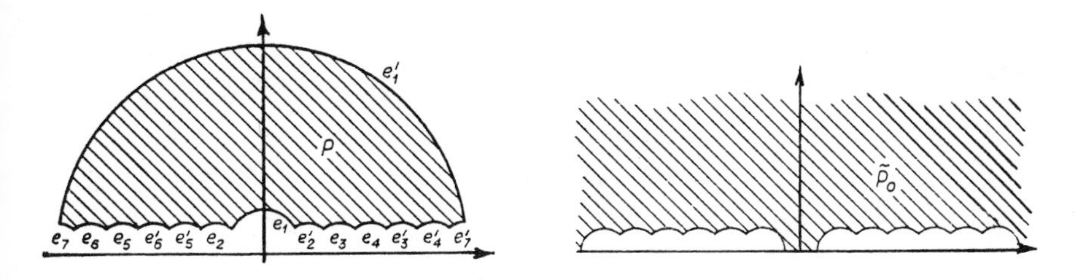

FIGURE 17

FIGURE 18

We choose the radii of the circles with centers on the real axis and the number p such that the conditions in the Poincaré theorem on a fundamental polygon of a Fuchsian group are satisfied (see Problem 9). Then the group G generated by linear fractional automorphisms of the upper half-plane that glue the sides of P in the order indicated in Figure 17 is a discrete and, consequently, a Fuchsian group of the first kind. Assume that it is generated by the elements $A_1, \ldots, A_n$, where $A_1(z) = 2z$.

Consider the subgroup G_0 of G generated by the elements $A_2, \ldots, A_n$. It is not hard to show that a fundamental polygon $\tilde{P}_0$ for G_0 can be obtained from P by removing the sides $|z| = 1$ and $|z| = 2$ and extending the other four sides (Figure 18). If we consider G_0 as a Kleinian group, then a fundamental region P_0 for G_0 is obtained by adjoining to $\tilde{P}_0$ the polygon symmetric to $\tilde{P}_0$ with respect to the real axis.

Let $f(z) = -2z$. It is easy to show that the group G_0 and the element f satisfy the conditions of the second Maskit combination theorem, and, consequently, the group $H = \langle G_0, f\rangle$ is Kleinian; a fundamental region for it is shown in Figure 19.

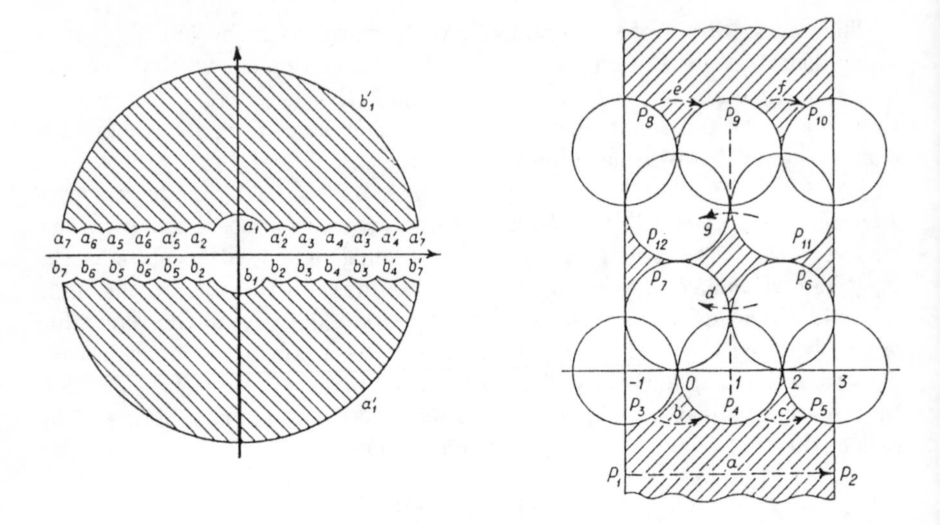

FIGURE 19 FIGURE 20

It is not hard to show that the set of discontinuity of H consists of two components which are carried one into the other by f. Then it follows from [92] that H is isomorphic to the fundamental group of a closed nonorientable surface.

EXAMPLE 28 (MASKIT [179]). *A Kleinian group isomorphic to the fundamental group of a closed orientable surface and not having invariant components.* Consider the linear fractional mappings given by the matrices

$$a = \begin{pmatrix} 1 & -4 \\ 0 & 1 \end{pmatrix}, \quad b = \begin{pmatrix} 1 & 0 \\ 1 & 1 \end{pmatrix}, \quad c = \begin{pmatrix} -3 & 4 \\ -1 & 1 \end{pmatrix},$$

$$d = \begin{pmatrix} -i & 2i \\ -1 & 2+i \end{pmatrix}, \quad e = \begin{pmatrix} 1+4i & 16 \\ 1 & 1-4i \end{pmatrix},$$

$$f = \begin{pmatrix} 3+4i & 12-16i \\ 1 & -1-4i \end{pmatrix}, \quad g = \begin{pmatrix} 3i & 8-6i \\ 1 & -2-3i \end{pmatrix},$$

and let G be the group generated by these mappings.

Let C_1 and C_2 be the respective lines $\operatorname{Re} z = 3$ and $\operatorname{Re} z = -1$, and let $C_3, \ldots, C_{12}$ be the circles of radius 1 about the respective points $-1, 1, 3, 2+i, i, -1+4i, 1+4i, 3+4i, 2+3i$ and $3i$. Denote by $P_1, \ldots, P_{12}$ the non-Euclidean planes in the half-space $\mathbf{R}^3_+$ with the circles $C_1, \ldots, C_{12}$ as their boundaries. Note that $a(P_1) = P_2, b(P_3) = P_4, c(P_4) = P_5, d(P_6) = P_7, e(P_8) = P_9, f(P_9) = P_{10}$, and $g(P_{11}) = P_{12}$. Partitioning each of the planes P_4 and P_9 into two parts (non-Euclidean) by a plane supported by the line $\operatorname{Re} z = 1$, we get a polyhedron Q whose sides are glued together pairwise by the mappings $a, \ldots, g$. It is not hard to see that any two sides of Q intersect at angles of either $\pi/2$ or π (Figure 20).

By the Poincaré theorem on a fundamental polyhedron (see Problem 9), Q is a fundamental polyhedron for G'. This implies that $\Omega(G)/G$ is the union of four spheres with three points deleted. All the defining relations in G have the form $cba = d^{-1}bdc = afe = g^{-1}egf = 1$. It is clear from this that G is generated by the mappings b, d, e, and g, and these generators satisfy the single relation

$$b^{-1}d^{-1}bdg^{-1}e^{-1}ge = 1, \tag{59}$$

i.e., G is isomorphic to the fundamental group of a closed orientable surface characterized by the relation (59).

§2. Limit sets of Kleinian groups

EXAMPLE 29 (ABIKOFF [2]). *A finitely generated Kleinian group with nonempty residual set.* Let Γ be a finitely generated Fuchsian group of the first kind with limit circle C, and suppose that the point at infinity is not a fixed point for elements of Γ. The isometric fundamental region $P(\Gamma)$ of Γ consists of two components: a bounded one $P_1 \subset \operatorname{int} C$ and an unbounded one P_2.

Take a hyperbolic or loxodromic transformation g with fixed points $\varsigma_1 \in P_1$ and $\varsigma_2 \in P_2$ such that its isometric circle $I(g)$ lies in P_1, while $I(g^{-1})$ lies in P_2. Then the combined group $G = \langle \Gamma, g \rangle$ is discontinuous, and the points ς_1 and ς_2 are in $\Lambda(G)$. However, these points do not fall on the boundary of any component of $\Omega(G)$ (so G is not a function group). The latter follows from the fact that the circles $g^n(C)$ and $g^{-n}(C)$, which lie entirely in $\Lambda(G)$, contract to ς_1 and ς_2, respectively, and, consequently, for any $z_0 \in \Omega(G)$ the circle $g^n(C)$ separates the points z_0 and ς_1 (and $g^{-n}(C)$ separates z_0 and ς_2) for sufficiently large n, i.e., neither ς_1 nor ς_2 can belong to the boundary of the component Ω_0 of $\Omega(G)$ containing z_0.

This example also carries over directly to the multi-dimensional case.

EXAMPLE 30. *Infinitely generated Kleinian groups whose residual limit sets have positive planar measure.* The group constructed in Example 33 has this property. We give another example, due to Abikoff [2]. Let C and C' be disjoint closed Jordan curves of positive area and obtained one from the other by a translation. Take four sequences $\{l_n\}, \{L_n\}, \{l'_n\}$ and $\{L'_n\}$ of disjoint Jordan curves distinct from C and C', with the first two sequences converging uniformly to C and the second two uniformly to C', and with

$$\begin{aligned} &l_n \subset \operatorname{int} l_{n+1} \subset \operatorname{int} C \subset \operatorname{int} L'_{n+1} \subset \operatorname{int} L_n, \\ &l'_n \subset \operatorname{int} l'_{n+1} \subset \operatorname{int} C' \subset \operatorname{int} L'_{n+1} \subset \operatorname{int} L'_n, \qquad n = 1, 2, \dots. \end{aligned} \tag{60}$$

We construct a Kleinian group G whose limit set contains all the curves in the indicated sequences (60). As in Example 18, for this it suffices to take a filling by disks U_k and U'_k of the regions bounded by these curves (excluding the region containing ∞) such that the disks accumulate at all the points of the curves, and to consider linear fractional mappings g_k having, together with g_k^{-1}, the corresponding circles ∂U_k and $\partial U'_k$ as their isometric circles. In a way analogous to that in the preceding example we get that $C \cup C' \subset \Lambda_0(G)$, and hence $m_2\Lambda_0(G) > 0$.

This example generalizes also to the case of dimension $n > 2$ if C and C' are taken to be closed Jordan surfaces of positive n-dimensional Lebesgue measure.

EXAMPLE 31 (BEARDON [42]). *A finitely generated Kleinian group in* $\mathbf{R}^n$, $n \geq 2$, *whose limit set has Hausdorff dimension greater than* $n/2$. Before presenting this example, we recall the concepts of Hausdorff dimension and Hausdorff measure for subsets of $\mathbf{R}^n$.

Let $h(r)$, $0 \leq r \leq r_0$, be a monotonically increasing function with $h(0) = 0$. Consider a set $E \subset \mathbf{R}^n$, $n \geq 1$, and let $\{E_k\}$ be a finite or countable covering of E by open sets E_k with diameters $d(E_k) = d_k$. Require that all the d_k be $< 2r$, and let

$$m_h(E; r) = \inf_{d_k < 2r} \sum_k h(d_k/2), \tag{61}$$

where the infimum is over all such coverings. The quantity (61) is a monotonically decreasing function of r and has a (finite or infinite) limit

$$m_h(E) = \lim_{r \to 0} m_h(E; r), \tag{62}$$

called the *Hausdorff h-measure of the set E*. In the case $h(r) = r^\alpha$ ($\alpha > 0$) formula (62) defines the so-called *Hausdorff measure of dimension α*. Moreover, for each compact set E there exists a unique nonnegative number $d(E)$ such that $m_{r^\alpha}(E) = \infty$ for $0 \leq \alpha < d(E)$ and $m_{r^\alpha}(E) = 0$ for $\alpha > d(E)$. This number $d(E)$ is called the *Hausdorff dimension of the set $E \subset \mathbf{R}^n$*.

We proceed to construct the example. Let G_k ($k = 1, 2, \dots$) be the group generated by the reflections with respect to the spheres of unit radius about the points $(a_1, \dots, a_n) \in \mathbf{R}^n$, where the a_j take the values $0, 3, 6, \dots, 6k$. Then $G_1 \subset G_2 \subset \cdots$; hence, $\Lambda(G_1) \subset \Lambda(G_2) \subset \cdots$. Let $d_k = d(\Lambda(G_k))$. Then all the d_k are $\leq n$, and the sequence d_k is monotonically increasing; consequently, $\lim_{k\to\infty} d_k = d \leq n$.

Consider now the subgroups G_k^0 of G_k of index 2 ($k = 1, 2, \ldots$). They have the same limit sets. It remains to prove that $d > n/2$, and then also $d_k > n/2$ beginning with some k. The proof of this can be found in [42].

In particular, the same construction can be applied also to the Schottky groups, and then we get a Schottky group G with $d(\Lambda(G)) > n/2$. This result was obtained in [24] for $n = 2$.

We remark that for these groups $G \subset \mathcal{M}_n$ and, what is more, for any geometrically finite Kleinian group $G \subset \mathcal{M}_n$ the Hausdorff dimensions $d(\Lambda(G))$ is always less than n; see Tukia [304]. At the same time, Sullivan [299] constructed an example of a finitely generated Kleinian group G_0 on the plane (geometrically infinite) such that $d(\Lambda(G_0)) = 2$.

EXAMPLE 32 (KRUSHKAL′ [131]). *Nonmeasurable sets—nonmeasurability of fundamental sets in the limit set* $\Lambda(G)$. Let G be a Kleinian group of Möbius automorphisms of the unit ball $B^n = \{x \in \mathbf{R}^n\colon |x| < 1\}$ in $\mathbf{R}^n$ ($n \geq 2$) that acts discontinuously in B^n and, possibly, on some set $\Omega \subset \partial B^n = S^{n-1}$ (which is open and dense in S^{n-1}); then $\Lambda(G) \subset S^{n-1}$. Denote by μ_{n-1} the $(n-1)$-dimensional spherical Lebesgue measure on S^{n-1}, and by $|A'(x)| = |dA(x)|/|dx|$ the linear dilation at the point x under a mapping $A \in G$.

If from the limit set $\Lambda(G)$ we remove the set F_0 of fixed points of elements of G (which has dimension at most $n-3$), then the remaining set $\Lambda_0 = \Lambda(G) \backslash F_0$ no longer contains points x such that $g_1(x) = g_2(x)$ for distinct $g_1, g_2 \in G$, i.e., any points in Λ_0 nonequivalent with respect to G have disjoint orbits. Using the axiom of choice, we can form a fundamental set e_Λ for G in Λ_0 that contains only one point from each orbit $G(x)$, $x \in \Lambda_0$. We have the following assertion:

If $\mu_{n-1}\Lambda(G) > 0$ *and*

$$\mu_{n-1}(\partial P(G) \cap \Lambda(G)) = 0 \tag{63}$$

for the isometric fundamental polyhedron $P(G)$ *of* G *in* $B \cap \Omega$, *then* e_Λ *is nonmeasurable* (*as is every subset* e' *of it with positive outer measure*).

Sullivan [228] showed that condition (63) is equivalent to the condition that the (hyperbolic) volume of the manifold B^n/G grow more slowly than the volume of the hyperbolic space $\mathbf{R}^n_+$ (for groups of the first kind).

In particular, this condition is satisfied automatically by groups such that $\overline{P}(G) \subset B^n$. Here is an elementary special case. Let G be a torsion-free Fuchsian group in B^2 with compact fundamental region (and, consequently, $\Lambda(G)$ is the circle $S^1 = \{z\colon |z| = 1\}$). By the stated theorem, e_Λ is not measurable on S^1.

We give a direct proof of this assertion. Assuming that G has a measurable fundamental set $e_\Lambda \subset S^1$, we can take a bounded measurable function $h(z) \not\equiv$ const on e_Λ and extend it automorphically to S^1 by the equalities $h(g(z)) = h(z), g \in G$. Then it is possible to construct a harmonic function $u(z)$ in the disk B^2 with $u(z)|_{S^1} = h(z)$. Being bounded, u takes the boundary values

$h(z)$ almost everywhere on S^1, along nontangential paths in general, and it is uniquely determined by these values. This gives us that $u(g(z))$ must equal $u(z)$ for all $g \in G$, and $u(z) \not\equiv$ const in B^2, which is impossible. Consequently, e_Λ is not measurable.

The proof of the general theorem was given in [131] and is based on other considerations.

REMARK. If $\Omega \neq \varnothing$, i.e., G acts discontinuously on S^{n-1}, then it is simpler to pass to the group $\alpha G \alpha^{-1}$, where α is a conformal mapping of B^n onto the half-space $\mathbf{R}^n_+$, and to consider instead of μ_{n-1} the usual Lebesgue measure m_{n-1} in $\mathbf{R}^{n-1}$ (and, in general, for $n > 2$ it is possible to start from a group that is discontinuous in $\mathbf{R}^{n-1}$).

EXAMPLE 33 (KRUSHKAL′). *Necessity of the spatial nature of condition* (63) *for nonmeasurability of a fundamental set in* $\Lambda(G)$. The example constructed below shows that for discontinuous groups in $\mathbf{R}^{n-1}$ ($n > 2$) condition (63) cannot be changed to the analogous condition on the boundary of the isometric fundamental region $P^0(G) = \overline{P}(G) \cap \Omega$ of the group G on S^{n-1} i.e., by the condition $\mu_{n-1}(\partial P^0(G) \cap \Lambda(G)) = 0$.

Indeed, let C be the circle $\{|z| = 1\}$, and let C_0 be a closed Jordan curve of positive area in the disk $U = \{|z| < 1\}$. As in Example 18, we construct a filling of the disk U by disks of pairwise equal radii such that these disks accumulate at C, do not intersect C_0, and accumulate also at C_0 from both sides, i.e., such that C_0 also belongs to the limit set of the group $G = \langle g_1, g_2, \dots \rangle$ obtained there. Then C is contained in $\partial P(G)$, but does not belong to the boundary of $P^0_G = \{z\colon |z| > 1\}$, and e_Λ is a measurable set of positive planar measure. We note also that here C is in the residual limit set of G.

EXAMPLE 34. *Fuchsian groups with measurable fundamental sets of positive measure in the limit set.* Such groups are obtained, in particular, from Examples 39 and 41. We present a simple modification of them for Fuchsian groups that gives a purely hyperbolic group. Let C be the unit circle, and C_1 and C_2 its upper and lower halves. On C_1 consider a Cantor set F_0 of positive linear measure with complementary open intervals δ_k ($k = 1, 2, \dots$), so that $F_0 \cup (\bigcup_k \delta_k) = C_1$. For each k consider the circle S_k orthogonal to C and such that $(\operatorname{int} C_k) \cap C = \delta_k$. Let F_0' be the symmetric image of F_0 on the semicircle C_2; the notation δ_k' and S_k' ($k = 1, 2, \dots$) has an analogous meaning.

For each pair of circles S_k and S_k' we take a hyperbolic linear fractional mapping γ_k which together with γ_k^{-1} has the circles S_k and S_k' as their isometric circles. Then the group $\Gamma = \langle \gamma_1, \gamma_2, \dots \rangle$ generated by these transformations is a Fuchsian group of the first kind, and the exterior of all the S_k in the unit disk U is the isometric fundamental polygon $P(\Gamma)$ of this group. It is clear that $\partial P(\Gamma) \cap C = F_0 \cup F_0'$, and F_0 serves as a partial fundamental set of the group Γ in C.

If F_0 is taken on a smaller arc of C_1, then we get a Fuchsian group of the second kind with analogous properties.

We present next an example due to Pommerenke [208]. Let Γ be a Fuchsian group of linear fractional automorphisms of the unit disk U. It is assumed that this group is of the so-called convergent type, i.e., such that

$$\sum_{\gamma\in\Gamma}(1-|\gamma(z)|^2)=(1-|z|^2)\sum_{\gamma\in\Gamma}|\gamma'(z)|<\infty, \qquad z\in U. \tag{64}$$

For such groups there exists a Green's function $g(z)$ defined as a Blaschke product associated with the points $\gamma(0), \gamma\in\Gamma$:

$$g(z)=\prod_{\gamma\in\Gamma}[e^{-i\theta(\gamma)}\gamma(z)] \qquad (\theta(\gamma)=\arg\gamma(0), \theta(I)=0); \tag{65}$$

in particular, $g(0)=0$ and $|g(z)|<1$. The function $g(z)$ has a nontangential limit $g(\varsigma)$ (i.e., a limit as $z\to\varsigma$ in angles with vertices $\varsigma\in\partial U$) with $|g(\varsigma)|=1$ for almost all $\varsigma\in\partial U$. It can be proved [208] that there exists a (maximal) region $D_g\subset U$ containing $z=0$ which the $g(z)$ in (65) maps univalently onto a star-shaped region in U; D_g is a fundamental region for a group Γ satisfying condition (64), and the planar measure of its boundary ∂D_g is equal to zero.

The set $A=\{\varsigma\in\partial U\colon |g(\varsigma)|=1, g'(\varsigma)\neq\infty \text{ exists}\}$ and the set $B=\{\varsigma\in\partial D_g\cap\partial U\colon |g(\varsigma)|=1\}$ are defined by means of the nontangential values $g(\varsigma)$ and $g'(\varsigma)$. It can be proved that B has positive logarithmic capacity. Define on ∂U the probability measure $\mu(E)=(1/2\pi)m_1(g(E\cap B))$ ($E\subset\partial U$ a Borel set); in particular, $\mu(\partial U)=1$. It turns out that for the symmetric difference of A and $\bigcup_{\gamma\in\Gamma}\gamma(B)$ this measure is equal to zero, and that its continuous part is

$$\mu_1(E)=\frac{1}{2\pi}\int_{E\cap A\cap B}|g'(z)|\,|dz|,$$

and $\mu_1(\partial U)=(1/2\pi)m_1(A)$. In the same place, in [208], conditions are written out that ensure that $m_1(A)>0$, which is equivalent to the existence of a normal fundamental region P of Γ with $m_1(\partial P\cap\partial U)>0$. From this, considering that the set $L=\partial P\cap\partial U$ is closed, we get the measurability of at least the partial fundamental set of Γ obtained by removing from L the union of the sides of $\overline{P}$.

§3. Quasiconformal deformations of Kleinian groups

EXAMPLE 35 (APANASOV [36], [38]). *A finitely generated quasi-Fuchsian group in space.* The properties of quasi-Fuchsian groups have been fairly thoroughly studied in the case of the plane. But in space even their existence was unknown until recently. Ahlfors and Krushkal′ have directed attention to this.

Suppose for simplicity that $n=3$, and consider the cube $Q=\{x\in\mathbf{R}^3\colon |x_i|<1/2\}$. Let S_i ($1\leq i\leq 8$) be spheres of radius $\sqrt{3}/3$ about the vertices of Q. They intersect at angles $\pi/3$ and are orthogonal to the sphere $S_0 = S(0, \sqrt{15}/6)$. Construct six more spheres S_i $(9 \leq i \leq 14)$ with radii $\sqrt{10}/6$ and with centers lying on the coordinate axes and at a distance $5/6$ from the origin. The spheres S_i $(9 \leq i \leq 14)$ are perpendicular to the spheres S_j $(0 \leq j \leq 8)$. Denote by L_i $(1 \leq i \leq 8)$ the planes passing through the points $e_3 = (0,0,1), (0,0,0)$, and the center of S_i. Further, let $L_9 = L_{10} = \{x\colon x_3 = 0\}$ and $L_{11} = L_{12} = L_{13} = L_{14} = \{x\colon x_1 = 0\}$. The group $G \subset \mathcal{M}_3$ generated by the involutions $g_i = O_i \circ J_i$, $1 \leq i \leq 14$, where J_i is inversion with respect to S_i and O_i is reflection with respect to the plane L_i, is a Fuchsian group of the first kind in $\mathbf{R}^3$; moreover, $\Lambda(G) = S_0 = S(0, \sqrt{15}/6)$.

If now the spheres S_{13} and S_{14} with centers $\pm(5/6)e_3$ (respectively, the spheres S_{11} and S_{12} or S_9 and S_{10}) are replaced by the spheres S_{13}^t and S_{14}^t with centers $\mp te_3$ and radii $\sqrt{t^2 - t + 5/12}$ (respectively, the spheres S_{11}^τ and S_{12}^τ or S_9^r and S_{10}^r with centers $\pm\tau e_2$ or $\pm re_1$ and analogous radii) and the group $G(r, \tau, t)$ generated by the involutions g_i with respect to the resulting system of spheres is constructed, then the group arising in this way is quasi-Fuchsian (when $r, \tau, t \in (5/6 - \varepsilon, 5/6 + \varepsilon)$ and $\varepsilon > 0$ is sufficiently small); in other words, there exists a quasiconformal automorphism f of $\overline{\mathbf{R}}^3$ such that $fGf^{-1} = G(r, \tau, t)$. Moreover, $f(S_0) = \tilde{S} = \Lambda(G(r, \tau, t))$ is a Jordan surface that does not have a tangent at any point. Its interior and exterior are invariant components of the set of discontinuity: $\Omega(G(r, \tau, t)) = \operatorname{ext} \tilde{S} \cup \operatorname{int} \tilde{S}$.

The quasiconformal deformation f is constructed by a certain geometric method that can be applied to many Fuchsian groups with compact fundamental polyhedra (see [36] and [37]).

EXAMPLE 36. *Infinitely generated quasi-Fuchsian groups that are not quasiconformal deformations of Fuchsian groups.* The construction in Example 37 is an example of such a group for $n = 2$; in that case the example gives a quasi-Fuchsian group with a fixed curve that has "null peaks". Consequently, such a curve cannot be a quasicircle, since angles are distorted by a finite factor under a quasiconformal mapping.

Another example of a quasi-Fuchsian group that is not a quasiconformal deformation of a Fuchsian group was constructed by Abikoff [2]. It has to do with another property of quasicircles, namely, the fact that they are Jordan curves with planar measure zero.

The construction of this example uses Osgood's [201] Jordan curve $C \subset \mathbf{C}$ of positive planar measure, which is the closure of a countable set of pairwise distinct linear segments I_k; the curve C becomes closed when two rectilinear segments are added.

We first construct a group G of mappings by choosing generating circles that satisfy the following conditions:

a) The centers lie in $\bigcup_k I_k$.

b) The interiors are disjoint.

c) Each generating circle is tangent precisely to two neighbors.

d) Except for its endpoints, each segment I_k is covered by the closed disks bounded by the generating circles with centers on I_k.

e) The radii of the generating circles tend to zero as the endpoints of I_k are approached.

f) Circles with centers on distinct I_k and I_m do not have common points.

Let G_0 be the subgroup of G of index 2 consisting of the orientation-preserving mappings. Then $\Lambda(G_0) = \Lambda(G)$; moreover, $\Lambda(G_0) \supset C \setminus \bigcup_k I_k$ and thus has positive measure. The group G is discontinuous, and it is necessary only to show that $\Lambda(G_0)$ is a Jordan curve. To do this, consider an arbitrary generating circle K with center on I_k. Since neighboring generating circles are tangent, $\Lambda(G_0)$ contains only two points of K, namely, the points a_k and b_k making up the intersection $K \cap I_k$. The points a and b corresponding to all the remaining circles after reflection with respect to K and with respect to the images lying inside K define a complete arc beteen a_k and b_k. This arc admits an (induced) parametrization and is easily shown not to have selfintersections.

EXAMPLE 37 (TETENOV [233]). *A quasi-Fuchsian group in space that is not a quasiconformal deformation of a Fuchsian group.* Consider the polycylinder

$$\begin{aligned} C = &\left\{ x \in \mathbf{R}^n \colon \sum_{k=1}^{n-1} x_k^2 = 1, x_n \geq 0 \right\} \\ &\cup \{ x \in \mathbf{R}^n \colon \sum_{k=1}^{n-1} x_k^2 \leq 1, x_n = 0 \}. \end{aligned} \tag{66}$$

For some ε $(0 < \varepsilon < 1/2)$ we take a system of disjoint spheres $\{S_i\}$ whose interiors cover the polycylinder (66) densely and which are orthogonal either to its base or to some generator.

Let the group $G \subset \mathcal{M}_n$ $(n \geq 3)$ consist of all possible superpositions of an even number of inversions with respect to the spheres S_i. This group is Kleinian. Its limits set $\Lambda(G), \mu_n(\Lambda(G)) = 0$, coincides with a certain Jordan surface lying in the ε-neighborhood of the polycylinder C, and separates the set $\Omega(G)$ of discontinuity of G into two invariant components, i.e., G is quasi-Fuchsian. It can be show [233] that there do not exist a Fuchsian group $\Gamma \subset \mathcal{M}_n$ and a quasiconformal automorphism f of $\overline{\mathbf{R}}^n$ such that $G = f\Gamma f^{-1}$, i.e., G is not a quasiconformal deformation of a Fuchsian group. The proof of this reduces to establishing that a component of the set of discontinuity (the exterior of the indicated Jordan surface) cannot be mapped quasiconformally onto a ball.

EXAMPLE 38 (APANASOV). *The connectedness of the isometric fundamental region is not invariant under quasiconformal deformations of groups.* We can take the group $\Gamma \subset \mathcal{M}_n$ $(n \geq 2)$ to be a group in $\overline{\mathbf{R}}^n$ with two generators $\gamma_1(x)$ and $\gamma_2(x)$ the same as the generators (49) in Example 19 (see Figure

13), while the remaining (hyperbolic) generators $\gamma_3, \gamma_4, \dots$ are such that Γ is the free product of the cyclic groups $\{\gamma_i\}_{i=1,2,\dots}$ and the limit set $\Lambda(\Gamma)$ coincides with the sphere S^{n-1}. Analogously, the group $G \subset \mathcal{M}_n$ ($G = f\Gamma f^{-1}$ in what follows) is taken to be a group with two generators $g_1(x)$ and $g_2(x)$ similar to the generators in (48) (see Figure 12), while the remaining ones are analogous to the generators $\{\gamma_m\}_{m=3,4,\dots}$ and are such that the limit points of the group $\prod_3^\infty\{\gamma_m\}$ lie on some Jordan surface bounding a simply connected region.

It is easy to see that the isometric polyhedron $P(\Gamma)$ always has two simply connected components, but the bounded component of the polyhedron $P(G)$ has a doubly generated fundamental group when $n \geq 3$, while $P(G)$ has three such components when $n = 2$.

However, if we consider fundamental regions $F(\Gamma)$ and $F(G)$ for Γ and G that are analogous to the regions (50) and (51), along with a quasiconformal mapping $f_1\colon F(\Gamma) \to F(G)$ carrying the sides of $F(\Gamma)$ into the corresponding sides of $F(G)$, then we can construct (by extending f_1 "by periodicity" and taking into account the removability of the limit sphere $\Lambda(\Gamma)$) a quasiconformal automorphism f of $\overline{\mathbf{R}}^n$ that is a quasiconformal deformation of Γ. Here $G = f\Gamma f^{-1}$.

EXAMPLE 39 (KRUSHKAL′). *A Kleinian group G on the plane which admits quasiconformal deformations that are conformal in $\Omega(G)$ and do not reduce to conformal mappings in the large.* As in Example 18, suppose that C is a closed Jordan curve in $\overline{\mathbf{C}}$, but now assume that it has positive planar measure. Using again a construction of a filling of the interior of C by disks of pairwise equal radii, we obtain a Kleinian group G such that $C \subset \Lambda(G)$. Now take an arbitrary measurable function $\mu(z) \not\equiv 0$ on C with $\|\mu\|_\infty < 1$ and extend it to the whole of $\Lambda(G)$ by the equalities $\mu(g(z)) = \mu(z)g'(z)/\overline{g'(z)}$ for $g \in G$ and $z \in C$, and $\mu(z) = 0$ for $z \in \Lambda(G)\backslash G(C)$. Then $\mu(z)\,d\bar{z}/dz$ is a Beltrami differential with respect to G that is concentrated only on $G(C)$. Further, setting $\mu(z) = 0$ in $\Omega(G)$, we obtain a quasiconformal automorphism $f^\mu\colon \overline{\mathbf{C}} \to \overline{\mathbf{C}}$ that is compatible with G, is conformal outside C, and does not reduce to a conformal automorphism in the large in $\overline{\mathbf{C}}$.

Another example of a nonrigid Kleinian group in $\overline{\mathbf{C}}$ is obtained from the construction in Example 41, which is based on other considerations.

These examples show that the known question of whether quasiconformal automorphisms of $\overline{\mathbf{R}}^n$ ($n \geq 2$) compatible with a group G reduce to conformal mappings on $\Lambda(G)$ (rigidity of the group) has a negative answer in the general case. It was already mentioned above in §2 for groups such that the condition (63) is satisfied.

EXAMPLE 40 (APANASOV [33]). *A "nonrigid" discrete group in $\mathbf{R}^n_+$, $n \geq 2$.* As has already been mentioned, the well-known Mostow rigidity theorem asserts, in particular, that if $G, G' \subset \mathcal{M}_n$ ($n \geq 3$) are Fuchsian groups in the ball B^n with fundamental polyhedra of finite hyperbolic volume, then any isomorphism of these groups of the form $g \to fgf^{-1} = g'$ generated by a quasiconformal automorphism f of B^n can be realized in the form $g \to AgA^{-1} = g'$, where $A \in \mathcal{M}_n$ and $A(B^n) = B^n$. There was a question [135]: Is it possible to get rid of the finiteness condition for the hyperbolic volume in this theorem, or, more precisely, are all Fuchsian groups of the first kind in $\mathbf{R}^n$ $(n \geq 3)$ rigid? The corresponding conjecture was suggested by nonmeasurability effects (see Example 32). The following example shows that the question has a negative answer in the general case.

Let Q be the $(n-1)$-dimensional unit cube, and let $\{k_m\}$ be a sequence of odd integers. We partition Q into $k_1^{(n-1)}$ equal cubes Q_{1i} and remove the interior of the central cube. At the second step we partition each of the remaining cubes Q_{1i} into $k_2^{(n-1)}$ cubes Q_{2j} and remove from them the interiors of the central cubes. This process is continued ad infinitum. By suitably choosing the numbers k_m (for example, $k_m = 3^m$) it is possible to make the $(n-1)$-dimensional measure $\mu_{n-1}(K)$ of the continuum K left in the limit equal to an arbitrary number between 0 and 1.

Consider the projection Δ_1 of the set $\Delta = Q/K$ on the first coordinate axis, and define a mapping $f\colon Q \times \mathbf{R} \to \mathbf{R}^n$ by the formulas

$$\begin{gathered} f(x) = (f_1(x), \dots, f_n(x)), \qquad f_1(x) = \int_0^{x_1} \varphi(y)\,dy, \\ f_i(x) = x_i, \qquad i = 2, \dots, n, \end{gathered} \tag{67}$$

where $\varphi(y) = 1$ for $y \in \Delta_1$ and $\varphi(y) = m > 1$ for $y \in [-1/2, 1/2] \backslash \Delta_1$. This mapping is quasiconformal, with $K_f = m$.

Consider now a system of pairwise disjoint $(n-1)$-dimensional balls $B_k \subset \Delta$ densely filling Δ, and let $\{S_k\}$ be a family of $(n-1)$-dimensional spheres in $\mathbf{R}^n$ orthogonal to the subspace $\mathbf{R}^{n-1}$ and such that $S_k \cap \mathbf{R}^{n-1} = \partial B_k$. Further, let $\{L_k\}$ be the family of $(n-1)$-dimensional planes passing through the centers of the spheres S_k and orthogonal to $\mathbf{R}^{n-1}$.

Similarly, we consider the system $\{S'_k, L'_k\}$ of spheres and planes corresponding to the system $\{S_k, L_k\}$ under the mapping f, and we define two Kleinian groups G and G' leaving $\mathbf{R}^n$ invariant. The group G is generated by the mappings

$$\begin{gathered} T_1(x) = x + e_1, \dots, T_{n-1}(x) = x + e_{n-1}; \\ g_j(x) = O_j \circ J_j(x), \qquad j = 1, 2, \dots . \end{gathered} \tag{68}$$

Correspondingly, the mappings

$$\begin{gathered} T'_x(x) = x + m_1 \cdot e_1, \quad T'_2(x) = x + e_2, \dots, T'_{n-1}(x) = x + e_{n-1}, \\ g'_j = O'_j \circ J'_j(x), \qquad j = 1, 2, \dots, \end{gathered} \tag{69}$$

are generators of G'; here O_j and O'_j are the reflections with respect to the planes L_j and L'_j, while J_j and J'_j are the inversions with respect to the spheres S_j and S'_j, and $m_1 = m_1(m)$ is a number greater than 1.

The mapping f given by (67) carries a fundamental polyhedron $F(G)$ of G into a polyhedron $F(G')$ in such a way that faces go into faces and vertices into vertices. Therefore, f can be extended from $F(G)$ to a mapping $\Phi\colon \mathbf{R}^n_+ \to \mathbf{R}^n_+$ by the rule

$$\begin{aligned} \Phi|_{F(G)} = f; \qquad \Phi \circ T_i(x) &= T'_i \circ \Phi(x), \qquad i = 1, \dots, n-1; \\ \Phi \circ g_j(x) &= g'_j \circ \Phi(x), \qquad j = 1, 2, \dots, \end{aligned} \tag{70}$$

where T_i, g_j and T'_i, g'_j are defined by (68) and (69).

By (70), the quasiconformal automorphism Φ of $\overline{\mathbf{R}}^n$ thus obtained is compatible with the action of $G, G' = \Phi G \Phi^{-1}$, and at the same time, it is clear that $G' \neq AGA^{-1}$ for all $A \in \mathcal{M}_n$, i.e., the quasiconformal deformation Φ is nontrivial.

We remark that the main role in our arguments is played by the fact that the $(n-1)$-dimensional measure of the set of limit vertices of the fundamental polyhedron $F(G)$ is positive. As established recently by Sullivan [228], this is a determining property for nonrigid groups.

EXAMPLE 41 (APANASOV [37]). *Kleinian groups G in space admitting quasiconformal deformations that are conformal in $\Omega(G)$ and do not reduce to conformal mappings in the large.* To construct such a group G we consider the construction in the preceding example and make the following modification in it: From the system of balls B_k densely filling the complement Δ of the continuum K in the cube Q we eliminate some ball B_0, and in the system of balls B'_k we eliminate the corresponding ball B'_0. Then not only are the resulting groups G and G' with generators (68) and (69) discrete, but their restrictions to $\mathbf{R}^{n-1}$ are Kleinian. Moreover, formulas (67) and (70) give a nontrivial quasiconformal deformation $\Phi\colon \overline{\mathbf{R}}^n \to \overline{\mathbf{R}}^n, G' = \Phi G \Phi^{-1}$, that obviously is conformal on $\Omega(G) = G(B_0)$ but does not reduce to a conformal mapping in the large; there is no $A \in \mathcal{M}_{n-1}$ such that $AGA^{-1} = G'$.

In this example the components $g(B_0), g \in G$, of the set $\Omega(G)$ are atoms. But it is not hard to construct nonrigid groups for which $\Omega(G)/G$ is an arbitrary closed hyperbolic manifold. To do this it is necessary to combine (freely) the group G constructed above with a corresponding Fuchsian group G_1 of the first kind such that the complement of the unbounded component of $P(G_1) \supset B_1$ lies in B_0, and the manifold B_1/G_1 is the desired one; then it is necessary to augment the system $\{B_k\}$ by balls exhausting $B_0 \backslash P(G_1)$. The nonrigid group obtained as a result has the indicated property.

EXAMPLE 42 (TETENOV [233]). *A quasiconformal deformation of a Fuchsian group can be a "cube group".* By a cube group we understand here a quasi-Fuchsian group whose limit surface approximates the boundary of a cube to a certain accuracy. One such group was constructed in [38]. We now construct a quasiconformal deformation of it into a Fuchsian group. For simplicity of notation we confine ourselves to 3-dimensional space.

Let $Q = \{x \in \mathbf{R}^3: |x_i| < 1/2\}$, and consider spheres S_k $(1 \le k \le 8)$ of radius $1/2$ about the vertices of Q. The intersections of these spheres with the faces of Q give arc quadrangles, and in each of them we inscribe a circle and extend it to a sphere with the same center. Continuing this process ad infinitum, we get a system of spheres S_k $(k = 1, 2, \ldots)$ mutually tangent from the outside and with centers on the faces of Q; the interiors of these spheres cover densely the surface of Q. Consider the group $\tilde{G}_1$ generated by the inversions J_k with respect to the spheres S_k, and denote by G_1 the subgroup of it of index 2 that consists of superpositions of an even number of such inversions. This group G_1 is quasi-Fuchsian. Its fundamental region can be taken to be the set

$$F(G_1) = \operatorname{int}\left\{\left(\bigcap_{k=1}^{\infty} \operatorname{ext} S_k\right) \cup J_1\left(\bigcap_{k=1}^{\infty} \operatorname{ext} S_k\right)\right\}, \tag{71}$$

which splits into two spherical polyhedra: $F_0 \ni 0$ and $F_1 \ni \infty$. The set of discontinuity of G_1 consists of two invariant components: $\Omega(G_1) = \Omega_0 \cup \Omega_1$, where

$$\Omega_0 = \bigcup_{g \in G_1} g(F_0), \qquad \Omega_1 = \bigcup_{g \in G_1} g(F_1). \tag{72}$$

Moreover, the component Ω_0 given in (72) lies interior to the cube Q and has a countable set of boundary circles lying on the boundary of Q.

To construct a Fuchsian group $G \subset \mathcal{M}_3$ quasiconformally equivalent to G_1 we consider a sphere S_0 orthogonal to the spheres $S_i' = S_i$, $i = 1, \ldots, 8$. The intersections of their exteriors with S_0 give spherical quadrangles, and in each of them we inscribe a circle, and so on; we construct a family $\{S_k'\}$ of spheres orthogonal to S_0 that is analogous to the family $\{S_k\}$. As above, we construct from this family a Kleinian group G that is a Fuchsian group of the first kind with limit sphere S_0.

To construct a quasiconformal deformation $\overline{\Phi}: \overline{\mathbf{R}}^3 \to \mathbf{R}^3, G_1 = \Phi G \Phi^{-1}$, it is necessary to quasiconformally map a fundamental polyhedron $F(G)$ onto the fundamental polyhedron $F(G_1)$ given by (71) with sides and faces corresponding. This can be done (see [233]) by means of a mapping that is the identity outside some neighborhood of $\Lambda(G_1)$, while on the rest of $F(G)$ it is a dilation along circles orthogonal to S_0. If we then extend this mapping in a way similar to that in the formulas (70) in Example 40, we obtain the desired quasiconformal deformation Φ.

EXAMPLE 43 (JØRGENSEN AND MARDEN [112]). *Topologically equivalent discrete groups that are not quasiconformally equivalent.* In this example we construct two discrete groups G_1 and G_2 of Möbius automorphsims of $\overline{\mathbf{C}}$ that have the following properties:

a) G_1 and G_2 are each freely generated by two elements with parabolic commutator.

b) G_1 and G_2 act discontinuously in $\mathbf{R}^3_+$, but are not discontinuous in $\overline{\mathbf{C}}$.

c) Their fundamental polyhedra have infinitely many faces.

d) The manifolds $\mathbf{R}^3_+/G_1$ and $\mathbf{R}^3_+/G_2$ are homeomorphic (to the product of an open interval and a punctured torus).

e) G_1 and G_2 are topologically equivalent, but not quasiconformally equivalent, i.e., they are conjugate in the group of topological automorphisms of $\mathbf{C}$ but not in the group of quasiconformal automorphisms of $\overline{\mathbf{C}}$.

Take G_1 to be the group generated by the two Möbius mappings γ_1 and γ_2 with matrices

$$\gamma_1 = \begin{pmatrix} c & c \\ c & 1 \end{pmatrix}, \qquad \gamma_2 = \begin{pmatrix} 2-2c & -1 \\ -1 & c \end{pmatrix},$$

where $c = 1/2 - i\sqrt{3}/2$. Then

$$\gamma_1\gamma_2 = \begin{pmatrix} 2-c & -1 \\ 1 & 0 \end{pmatrix}, \qquad \gamma_2\gamma_1 = \begin{pmatrix} 2-c & 1 \\ -1 & 0 \end{pmatrix},$$

$$\gamma_1\gamma_2^{-1} = \begin{pmatrix} -1+2c & 2+c \\ c & 2-c \end{pmatrix},$$

and $\gamma_1\gamma_2\gamma_1^{-1}\gamma_2^{-1}$ is the translation by $4-2c$. Denote this translation by g_1, and let g_2 be the translation by c.

Let Γ_1 be the group generated by g_1 and g_2. This group acts discontinuously in $\mathbf{R}^3_+$. The fundamental polyhedron of Γ_1 in $\mathbf{R}^3_+$ is taken to be an infinite parallelepiped whose faces are identified by means of g_1 and g_2.

We now construct a fundamental polyhedron P_1 of G_1 in $\mathbf{R}^3_+$. It will thereby be proved that G_1 acts discontinuously in $\mathbf{R}^3_+$. Consider all the mappings obtained from $\gamma_1, \gamma_2, \gamma_1^{-1}$ and γ_2^{-1} by conjugation with elements of Γ_1. It is not hard to verify that the intersection of the exteriors of their isometric spheres in $\mathbf{R}^3_+$ is an infinite hyperbolic convex polyhedron P whose faces are regular hyperbolic hexagons, and if some face of P lies on the isometric sphere of an element $\gamma \in G_1$, then its image under γ is a face of P lying on the isometric sphere of γ^{-1}. Orthogonal projection on the plane $\overline{\mathbf{C}}$ maps each face of P onto a regular hexagon in $\overline{\mathbf{C}}$, and the projection of the whole boundary of P thereby gives a paving with regular hexagons (see Figure 21, where A and B denote generators of G_1 conjugate to γ_1 and γ_2). It can be shown that the intersection of P with a standard polyhedron of the cyclic group $\langle g_1 \rangle$ (i.e., with a shell of thickness $2\sqrt{3} = |4-2c|$) gives a fundamental polyhedron P_1 of the group G_1. In Figure 21 the dotted lines show polygonal lines $\gamma_t, t \in \mathbf{R}$, on the boundary of P that are invariant under the action of the group $\langle g_1 \rangle$. By considering for $t \in \mathbf{R}$ the sections of P_1 of the form

$$S_t = \{x = (z, x_3) \in \mathbf{R}^3_+ : z \in \gamma_t,\ x_3 \geq x_3(z)\}$$

and observing that the surfaces S_t/G_1 are homeomorphic to a punctured torus it is possible to prove that $\mathbf{R}^3_+/G_1$ is homeomorphic to the product of an open interval and a punctured torus.

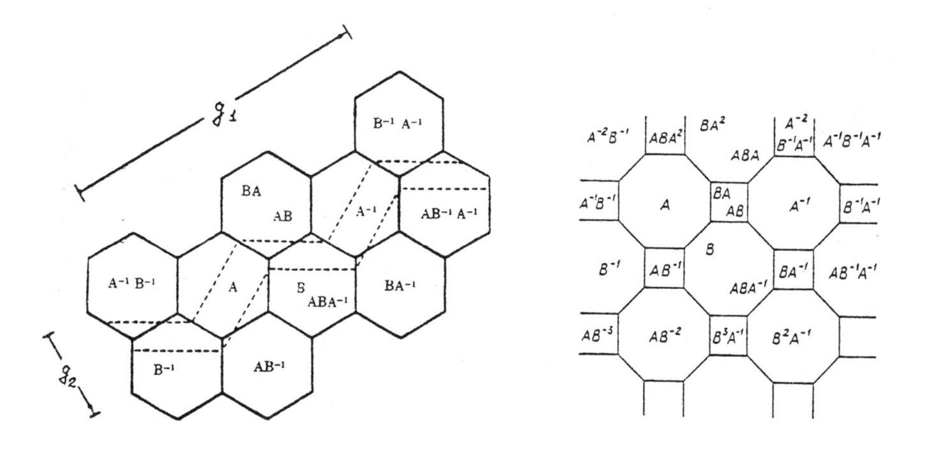

FIGURE 21 FIGURE 22

We now take G_2 to be the group generated by the two Möbius mappings γ_3 and γ_4 with matrices

$$\gamma_3 = a^{-1}\begin{pmatrix} 1 & -i \\ 1 & 1 \end{pmatrix}, \qquad \gamma_4 = b^{-1}\begin{pmatrix} 2+i & i \\ -1 & -i \end{pmatrix},$$

where $a = \sqrt{1+i}$ and $b = \sqrt{1-i}$. Then

$$\gamma_3\gamma_4 = ab^{-1}\begin{pmatrix} 2 & i \\ 1 & 0 \end{pmatrix}, \qquad \gamma_4\gamma_3 = ab^{-1}\begin{pmatrix} 2 & -i \\ -1 & 0 \end{pmatrix},$$

and, consequently, the commutator $g_3 = \gamma_3\gamma_4\gamma_3^{-1}\gamma_4^{-1}$ is the translation $z \to z+4$. Let $g_4(z) = z - 2i$, and let $\Gamma_2 = \langle g_3, g_4 \rangle$.

A fundamental polyhedron P_2 of G_2 is the intersection of the layer $\{z = x+iy\colon a < y < a+2\}$ and the common exterior P' of the isometric spheres in R^3_+ of the elements that are conjugate to $\gamma_3, \gamma_4, \gamma_3^{-1}, \gamma_4^{-1}$ by elements of Γ_2. The boundary of P' consists of a union of hyperbolic squares and regular octagons; its orthogonal projection on $\overline{\mathbf{C}}$ is shown in Figure 22, where A and B denote generators of G_2 conjugate to γ_3 and γ_4.

The manifold $\mathbf{R}^3_+/G_2$ is also homeomorphic to the product of an open interval and a punctured torus; hence G_1 and G_2 are topologically equivalent. However, it can be shown that these groups are not quasiconformally equivalent; see [112] for details.

§4. Kleinian groups that are not geometrically finite

EXAMPLE 44 (BERS [51]) *Degenerate groups.* Recall that degenerate groups are characterized by the following properties: they are nonelementary and finitely generated, and their sets of discontinuity are contractible regions, i.e., have trivial homotopy groups. Up to this time there are no constructive examples of such groups, and this question remains one of the unsolved problems in the theory of Kleinian groups. At present there is only one proof of the existence of such groups, in the planar case (due to Bers). It is based on properties of Teichmüller spaces and dimension theory arguments. Degenerate groups provided the first example of groups that are not geometrically finite; this was discovered by Greenberg [87].

Extending a degenerate group from the plane to the space $\mathbf{R}^n$ $(n \geq 3)$, we certainly obtain a spatial degenerate group. However, it is not clear if there exist inherently spatial degenerate groups (for example, groups for which the dimension of the limit set is greater than 1).

As Marden showed [168], Klein-Maskit combination of degenerate groups with any b-groups gives again Kleinian groups that are not geometrically finite.

EXAMPLE 45 (ACCOLA [9]). *A "degenerate" (but infinitely generated) Kleinian group.* This example gives an infinitely generated group whose set of discontinuity is a contractible region.

By induction we construct a sequence $\{a_n\}$ of numbers $a_n > 0$ such that the circles

$$\begin{gathered} S_{n,k} = \{z \in \mathbf{C}\colon |z - k/2^n - ia_n| = a_n\}, \\ k = 1, 3, \dots, 2^n - 1, \ n = 1, 2, \dots, \end{gathered} \tag{73}$$

are disjoint. It is clear that $a_n \to 0$. Denote by $\{S'_{n,k}\}$ the system of circles that are the mirror reflections of the circles $S_{n,k}$ with respect to the real axis $\mathbf{R}$.

Consider the group G of linear fractional mappings that is generated by the parabolic transformations $g_{n,k} = O \cdot J_{n,k}$, where $J_{n,k}$ is inversion with respect to the circle $S_{n,k}$, and O is reflection with respect to the axis $\mathbf{R}$. The group G is Kleinian. Its isometric fundamental region

$$P(G) = \bigcap_{n,k=1}^{\infty} \operatorname{ext}(S_{n,k} \cup S'_{n,k})$$

consists of a single connected component. Moreover, any image $g_{n,k}(\infty)$ can be joined to ∞ by a path lying in $\Omega(G)$. Therefore, $\Omega(G)$ is connected. At the same time, $P(G)$ is simply connected, and $\Omega(G) = \bigcup g_{n,k}(P(G))$.([10]) Consequently, $\Omega(G)$ is simply connected. Thus, the limit set $\Lambda(G)$ of G is a continuum containing the interval $[0, 1]$ of the real axis.

We note that the above construction carries over immediately to any dimension $n > 2$ (with the circles (73) replaced by corresponding spheres);

([10]) *Translation editor's note*: The union should be taken over all elements of G.

moreover, the limit set of the group obtained can be made to contain any surface homeomorphic to the ball B^k of dimension $k \leq n-1$.

EXAMPLE 46 (RILEY [97, P. 253]). *A limit approach which gives a discrete group that is not geometrically finite.* Let Γ_α be the group of linear fractional transformations on $\overline{\mathbf{C}}$ generated by the mappings

$$\gamma_1(z) = z+1, \qquad \gamma_\alpha(z) = z/(\alpha z+1).$$

The isometric circles $I(\gamma_\alpha)$ and $I(\alpha_\alpha^{-1})$ of the second mapping are tangent at zero and have radius $1/|\alpha|$. Therefore, for large $|\alpha| \geq 4$ the group Γ_α is a free geometrically finite Kleinian group. On the other hand, Γ_α is not discrete for $\alpha \neq 0$, $|\alpha| < 1$, because it has elements with isometric circles of radius greater than 1 (see, for example, [149]).

Now let the domain $D \subset \mathbf{C}$ consist of the points α for which Γ_α has the following properties: 1) Γ_α is discrete; 2) Γ_α is geometrically finite; 3) Γ_α is freely generated by $\gamma_1(z)$ and $\gamma_\alpha(z)$; and 4) all the parabolic elements in Γ_α are conjugate by powers of $\gamma_1(z)$ and $\gamma_\alpha(z)$. Since D lies outside the unit disk, its boundary ∂D contains a continuum. Therefore, there exists a boundary point $\beta \in \partial D$ such that Γ_β has property 4). This group Γ_α is a limit (with respect to generators) of free Kleinian groups $\Gamma^\alpha, \alpha \in D$, satisfying 1)–4), and is thus discrete and free. However, property 3) of the limit group Γ_β contradicts its geometric finiteness [168].

We remark that a more explicit construction of doubly generated groups with similar properties is obtained in Examples 43 and 53.

EXAMPLE 47. *Kleinian groups that are not quasiconformally stable.* The class of such groups was first mentioned by Bers [51]. They are the degenerate groups corresponding to boundary points of the Teichmüller space of a finitely generated Fuchsian group of the first kind. For each such degenerate group G there exists an arbitrarily small admissible deformation carrying this group into a quasi-Fuchsian group. Namely, for G there exists a sequence of quasi-Fuchsian groups G_n isomorphic to it which converge to G in the sense of convergence on generators. It is clear that the isomorphism $G \cong G_n$, even if it preserves the type of the elements, cannot be induced by a quasiconformal automorphism of the plane, because G and G_n have a different number of components.

The following example is due to Yu. S. Il′yashenko. It is based on the result given above and the Maskit combination theorem. In passing we use the example to explain once more the geometric sense of the combination theorems.

Take a degenerate group G_1 on the boundary of the Teichmüller space $T_{1,1}$ of punctured tori along with a small admissible deformation $\chi\colon G_1 \to G_1'$, where G_1' is quasi-Fuchsian; then the quotient $\Omega(G_1)/G_1 = S_0'$ for G_1 (and it is also one of the quotients for G_1') is a punctured torus. Suppose that the loop $\alpha \subset S_0'$ is a topological circle going around the deleted point, and that one of the transformations corresponding to α in the groups G_1 and G_1' is the translation $z \to z+1$. Assume further that the invariant component Δ_{G_1} of G_1 (which coincides with its set $\Omega(G_1)$ of discontinuity) contains the half-plane $\operatorname{Im} z \leq 0$ (all these requirements can be satisfied by using a lemma of Ahlfors [15] and replacing G_1, if necessary, by a conjugate group: see Exercise 25). The same thing is assumed for G_1'.

Now take a quasi-Fuchsian (or even Fuchsian) group $G_2 \in T_{1,1}$ that represents two punctured tori $S_0'' = \Delta_1/G_2$ and $S_2 = \Delta_2/G_2$, where Δ_1 and Δ_2 are (invariant) components of $\Omega(G_2)$, and assume that the transformation $z \to z+1$ corresponds to a loop $\alpha_1 \subset S_0''$ going around the deleted point there, while the component Δ_1 contains the half-plane $\operatorname{Im} z \geq 0$. Then it is possible to use the Maskit combination theorem, and we obtain Kleinian groups $G = \langle G_1, G_2 \rangle$ and $G' = \langle G_1', G_2' \rangle$ which are b-groups. The group G is partially degenerate, but G' is nondegenerate. The quotient $\Omega(G')/G'$ consists of three surfaces: S_0' (a closed surface of genus 2)([11]) , determined by invariant component of a group G_0, and two punctured tori obtained by contracting S_0' along some loop $\tilde{\alpha}'$ on it; $\Omega(G)/G$ consists of a surface S_0 homeomorphic to S_0', and a punctured torus obtained from S_0 by contracting along some loop $\tilde{\alpha} \subset S_0$. Consequently, G and G' are not quasiconformally equivalent.

We now show that G and G' are connected by an admissible homomorphism close to the identity. Let us extend the isomorphism $\chi\colon G_1 \to G_1'$ to an isomorphism $G \to G'$ (denoted by the same letter) that is the identity on G_2. This isomorphism is still admissible. Indeed, the types of the transformations $g \in G$ and $\chi(g) \in G'$ are determined by the corresponding elements in the fundamental group of the surface S_0 (respectively, S_0'); namely, there are natural isomorphisms $G \xrightarrow{\pi} \pi_1(S_0)$ and $G' \xrightarrow{\pi'} \pi_1(S_0')$. The groups G and G' do not contain elliptic elements. An element $g \in G$ is parabolic if and only if $\pi(g) \in \pi_1(S_0)$ is an element conjugate to the loop $\tilde{\alpha}$. But $\pi(g)$ is conjugate to $\tilde{\alpha}$ if and only if $\pi'(\chi(g))$ is conjugate to the loop $\tilde{\alpha}'$, to which in G' there corresponds a basis in the set of conjugate accidental([12]) parabolic elements. This shows that the isomorphism χ is admissible, and, consequently, G is not quasiconformally stable.

EXAMPLE 48 (MASKIT [185]). *A web group that cannot be constructed from elementary groups.* Maskit [185] constructed the following interesting example of a web group. Let H and H^* be finitely generated purely hyperbolic Fuchsian groups of the first kind acting in the unit disk U. Consider a transformation $a_t(z) = tz$, where $t \in \mathbf{C}$, and let $H_t = a_t H a_t^{-1}$. Let G_t be the group generated by H_t and H^*. For sufficiently small $|t|$ the group G_t can be formed from H_t and H^* by Klein combination; for such t it is the free product of H_t and H^* and does not contain parabolic elements.

([11]) *Translation editor's note*: S_0' here is different from S_0' above.

([12]) These are the parabolic elements of G' to which there correspond hyperbolic elements in a Fuchsian equivalent $\Gamma' = hG'h^{-1}$, where h is a conformal mapping of an invariant component $\Delta_{G'} \subset \Omega(G')$ onto the disk U.

It can be shown that there exists a ray $\arg t = \theta_0$ on which every element of G_t is loxodromic. Assume that $\theta = 0$, i.e., t is a real number. Let T be the set of all t such that G_t can be formed by free combination in the Maskit sense. It follows from the first combination theorem that T is open. Let $t_0 = \sup T$ and let $G = G_{t_0}$. Then G is a purely loxodromic group that is the free product of H^* and H_{t_0}, and a fundamental set of it is the union of a fundamental set of H^* in the region $\{|z| > 1\}$ and a fundamental set of H_{t_0} in the disk $\{|z| < t_0\}$. It can be shown that any component of G is conjugate to either $\{|z| < t_0\}$ or $\{|z| > 1\}$. This gives us that the stabilizer of each component of G is a Fuchsian group of the first kind, i.e., that G is a web group.

Abikoff conjectured that the group constructed above cannot be constructed from elementary groups. A proof of this conjecture was given by Gusevskiĭ. It is based on the following arguments: If G were constructed in the Maskit sense from elementary groups, then a fundamental polyhedron of G would have to have finitely many faces. Since G does not contain parabolic elements, this implies that $M(G)$ is compact. Observing that G is a nontrivial free product and using Kneser's theorem (see §7 in Chapter II), we get that $M(G)$ contains a sphere which decomposes the fundamental group $\pi_1(M(G))$ into a free product; but this contradicts the fact that $M(G)$ is irreducible.

EXAMPLE 49 (ABIKOFF [3]). *A finitely generated Kleinian group whose residual limit set contains a circle.* Let G be the Kleinian group constructed in the preceding example, and let G' be the Kleinian group conjugate to G by reflection with respect to the unit circle. Using the first Maskit combination theorem again, we get that the group $\tilde{G} = \langle G, G' \rangle$ is Kleinian and is the free product of G and G' with an amalgamated Fuchsian subgroup H of the first kind. The limit set $\Lambda(H)$ has empty intersection with the boundary of each component of $\tilde{G}$ and thus lies in the residual limit set.

EXAMPLE 50 (ABIKOFF [3]). *A finitely generated Kleinian group whose residual limit set contains the limit set of a degenerate group.* Let G_0 be a degenerate group such that the point $z = 0$ lies in its set of discontinuity. Let G_1 be a purely hyperbolic Fuchsian group of the first kind, and let G_t be the conjugate of G_1 by the transformation $a_t(z) = tz$, $t \in \mathbf{C}$. For sufficiently small $|t|$ the group $G^t = \langle G_0, G_t \rangle$ is Kleinian and is the free product of G_0 and G_t. Just as in Example 48, it can be shown that there exists a ray $\arg t = \theta_0$ on which each element in G^t is loxodromic. Assume that $\theta_0 = 0$. Let T be the set of t such that the group G^t constructed from G_0 and G_t, with the help of the first combination theorem, is Kleinian. Then T is open. Let $G = G_{t_0}$, where $t_0 = \sup T$. It can be shown that G is Kleinian, and the limit set of the subgroup G_0 of it is contained in the residual limit set $\Lambda_0(G)$.

EXAMPLE 51 (ABIKOFF [3]). *A finitely generated discrete purely loxodromic group that is not discontinuous and has fundamental polyhedron with infinitely many faces.* Let G be the Kleinian group constructed in the preceding example, and denote by G' the group conjugate to G by reflection in the circle $|z| = t_0$. By the combination theorem, the group $H = \langle G, G' \rangle$ is the free product of G and G' with an amalgamated Fuchsian subgroup. It can be shown that: a) a fundamental set of H has empty interior; b) H is discrete; and c) a fundamental polyhedron of H has an infinite set of faces. These assertions are proved in [3]).

EXAMPLE 52 (APANASOV). *Infinitely generated Kleinian groups G for which $\Omega(G)/G = S$ is a surface of finite type.* Let $\mathcal{S}$ and $\mathcal{T}$ be two systems of circles (Figures 23 and 24). Denote by G_1 and $\tilde{G}_2$ Kleinian groups consisting of superpositions of an even number of inversions with respect to the systems $\mathcal{S}$ and $\mathcal{T}$, respectively. The discontinuity of the groups follows from use of the result in Problem 8 and consideration of the extended groups in the half-space $\mathbf{R}^3_+$. Denote now by G_2 the Kleinian group obtained from $\tilde{G}_2$ by adjoining two more generators, the translations $\gamma_1(z) = z + 1$ and $\gamma_2(z) = z + i$.

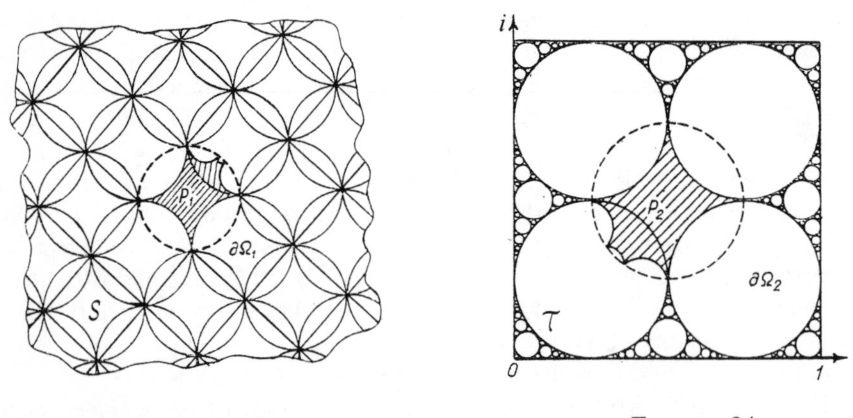

FIGURE 23 FIGURE 24

Fundamental polygons P_1 and P_2 for the groups G_1 and G_2 are shaded in Figures 23 and 24. They lie in the components $\Omega_i \subset \Omega(G_i)$; the $\partial\Omega_i$ are circles marked by dashes ($i = 1, 2$). All other components of G_1 and G_2 are conjugates of Ω_1 and Ω_2, respectively, and their stabilizers are 3-generated Fuchsian groups G_1^0 and G_2^0 of the first kind. Thus,

$$\Omega(G_1)/G_1 = \Omega_1/G_1^0 \approx \Omega_2/G_2^0 = \Omega(G_2)/G_2 \approx S^2 \backslash \{x_1, x_2, x_3, x_4\},$$

a sphere with four points deleted (here the sign $\approx$ indicates conformal equivalence).

We should point out the essential difference between the manifolds $(\mathbf{R}_3^+ \cup \Omega(G_1))/G_1$ and $(\mathbf{R}^3_+ \cup \Omega(G_2))/G_2$; their boundaries, as indicated, are conformally equivalent, but of the coverings corresponding to these manifolds the first is branched, while the second is not. Other groups with analogous properties are given in Example 41.

EXAMPLE 53. *Doubly generated Kleinian groups that are not geometrically finite.* The following example of Jørgensen [110] is interesting not only because a fundamental polyhedron of the group has countably many faces, but also because the manifold $M(G^*)$ is fibered over a circle. See Example 63 for details.

Consider for $m \geq 2$ the numbers $\lambda, \varphi, \rho, x$, and y given by the following equalities:

$$\begin{aligned} \lambda &= \exp(\pi i/2m), \\ 2\varphi &= 1 + (17 - 8\cos(\pi/m))^{1/2}, \\ 2\rho &= (\varphi+2)^{1/2} + (\varphi-2)^{1/2}, \\ 2x(\varphi-2)^{1/2} &= (3-\varphi)^{1/2} + (-\varphi-1)^{1/2}, \\ 2y(\varphi-2)^{1/2} &= -(3-\varphi)^{1/2} + (-\varphi-1)^{1/2}. \end{aligned} \tag{74}$$

Let G^* be the group generated by the matrices

$$T = \begin{pmatrix} \rho & 0 \\ 0 & \rho^{-1} \end{pmatrix}, \qquad X = \begin{pmatrix} -\lambda x & -(1+x^2) \\ 1 & \lambda^{-1}x \end{pmatrix}.$$

It can be assumed that G^* is a group of motions of the 3-dimensional hyperbolic space $\mathbf{R}^3_+$. Consider its commutator subgroup. This is the group $G \subset G^*$ generated by the transformations $Y = TX^{-1}T^{-1}X$ and X. Then the subgroup $\Gamma \subset G^*$ generated by the transformations T and

$$K = XYX^{-1}Y^{-1} = \begin{pmatrix} -\lambda^2 & 0 \\ 0 & -\lambda^{-2} \end{pmatrix}$$

is abelian and consists of Euclidean similarities. Let us consider the set

$$I = \{\gamma X\gamma^{-1}, \gamma Y\gamma^{-1}, \gamma X^{-1}\gamma^{-1}, \gamma Y^{-1}\gamma^{-1} : \gamma \in \Gamma\} \tag{75}$$

and take the non-Euclidean polyhedron $P \subset \mathbf{R}^3$ which is the exterior of all the isometric spheres of elements of I. Figure 25 shows the projection of part of these spheres on the boundary plane $\partial\mathbf{R}^3_+ = \mathbf{C}$. The polyhedron P is bounded by an infinite set of hyperbolically similar hexagons. We remark that the radii of the isometric spheres of the mappings YX, X, Y, and XY^{-1} are equal to $\rho^{-1}, 1, \rho$, and ρ^2, respectively, where ρ is determined by (74); by computations we can also get the parameters of the boundary hexagons—Figure 26 shows the projection of such a hexagon lying on the isometric sphere of X. If A and B denote some pair of generators of G taken from the set I and having commutator $ABA^{-1}B^{-1}$ equal to K, then the projection of P on the plane $\mathbf{C}$ is as in Figure 27.

FIGURE 25

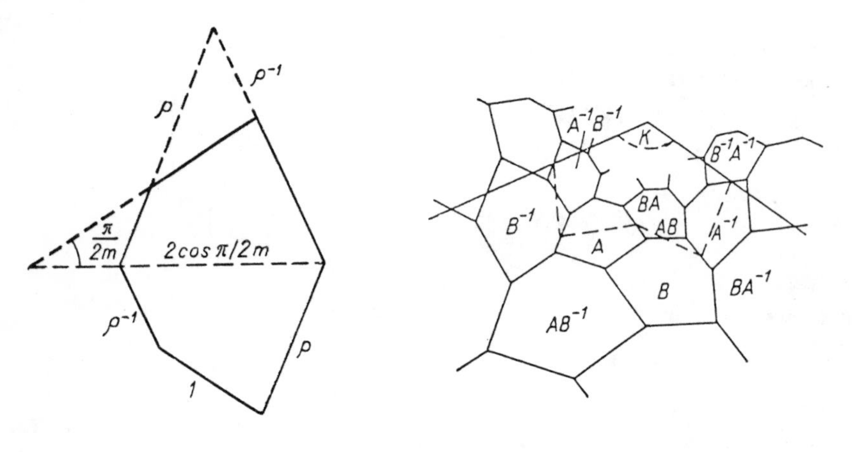

FIGURE 26 FIGURE 27

The subgroup $\Gamma \subset G^*$ acts on the boundary of P as a group of automorphisms. A fundamental polyhedron of G^* can be obtained from P as follows. Consider the intersection Q of P with the dihedral angle of opening $2\pi/m$ whose generators are two rays on $\mathbf{C}$ that are carried one into the other by the transformation K. We now take two hyperbolic planes (which are hemispheres about zero) that are carried one into the other by T, and consider the polyhedron R obtained by intersecting the region bounded by these planes with the polyhedron Q; and we have constructed a fundamental polyhedron for G^*. The proof of this fact can be obtained with the help of the Poincaré theorem (see Problem 9) by using the existing identification of sides of R. This gives us that the groups G^* and $G \subset G^*$ are discrete, while the polyhedron $Q = \bigcup T^n(R)$ is fundamental for G. It has infinitely many pairwise equivalent faces, of which two are planes orthogonal to $\mathbf{C}$, while the others are parts of isometric spheres of transformations in the set (75).

Two other examples of doubly generated groups that are not geometrically finite are given in §3 (see Example 43). They also provide discrete groups that are topologically but not quasiconformally equivalent. Figures 21 and 22 show the schemes of the projections of the fundamental polyhedra of these groups on the plane $\overline{\mathbf{C}} = \partial\mathbf{R}^3_+$.

§5. Uniformization: Hyperbolic manifolds

We first present some of the simplest (but important for what follows) examples of 3-dimensional manifolds that can be uniformized by Kleinian groups. These groups act discontinuously in $\mathbf{R}^3_+ \cup \overline{\mathbf{C}}$. We assume that $z = x_1 + ix_2$, and $x = (x_1, x_2, x_3) \in \mathbf{R}^3_+ \cup \overline{\mathbf{C}}$.

EXAMPLE 54. Let G be the cyclic group generated by the parabolic mapping $g\colon z \to z+1$. Then the set

$$F_G = \{(x_1, x_2, x_3)\colon 0 < x_1 < 1\} \tag{76}$$

is a fundamental polyhedron for G. If we identify the faces $L_1 = \{(x_1, x_2, x_3)\colon x_1 = 0\}$ and $L_2 = \{(x_1, x_2, x_3)\colon x_1 = 1\}$ of the polyhedron (76), then we obtain the manifold $M(G)$. This manifold is homeomorphic to $S^1 \times (0,1) \times [0,1) \equiv U' \times (0,1)$, where S^1 is the circle and $U' = \{z\colon 0 < |z| \le 1\}$ and is called a *punctured cylinder* (see Figure 28). The manifold $M(G)\backslash\partial M(G) = \mathbf{R}^3_+/G$ is hyperbolic.

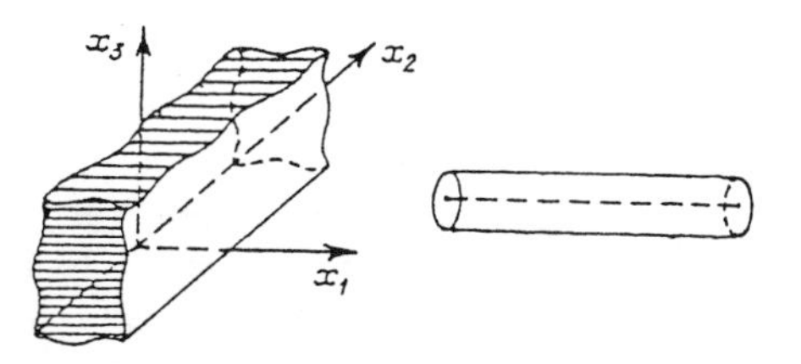

FIGURE 28

EXAMPLE 55. Let G be the cyclic group generated by a loxodromic mapping g that leaves $\mathbf{R}^3_+$ invariant and is such that the isometric spheres $I(g)$ and $I(g^{-1})$ are disjoint. Then the polycylinder bounded by them in $\mathbf{R}^3_+$ is a fundamental polyhedron for G. Identifying the faces of $I(g)$ and $I(g^{-1})$ of this polyhedron, we get a manifold that is homeomorphic to $S^1 \times \overline{U}$, where $\overline{U} = \{z\colon |z| \le 1\}$. This manifold is called a *solid torus* (see Figure 29).

Removal of the boundary $\partial M(G)$ yields a hyperbolic manifold $\mathbf{R}^3_+/G$ homeomorphic to $S^1 \times U$.

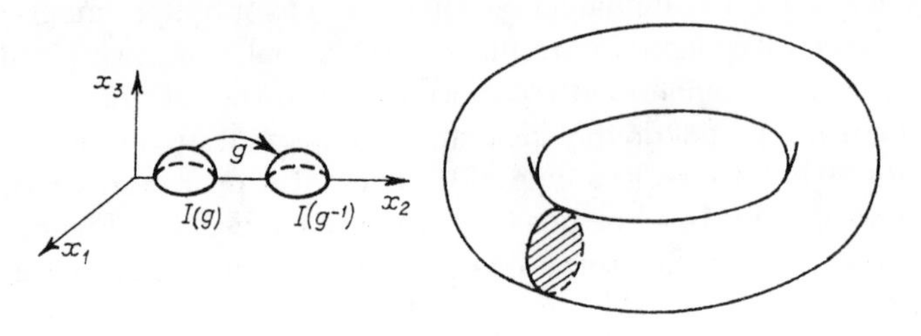

FIGURE 29

EXAMPLE 56. Let G be the free abelian group generated by the two parabolic mappings $g_1(z) = z + 1$ and $g_2(z) = z + i$. Then the set $F_G = \{(x_1, x_2, x_3): 0 < x_1 < 1, 0 < x_2 < 1\}$ is a fundamental polyhedron for G. If we identify the faces of this polyhedron, then we get a manifold $M(G)$ that is homeomorphic to $S^1 \times S^1 \times [0, 1) = U' \times S^1$. This manifold is called a *punctured torus* (see Figure 30); it is turned into a hyperbolic manifold by removing the boundary $\partial M(G)$, which is homeomorphic to the torus $S^1 \times S^1$.

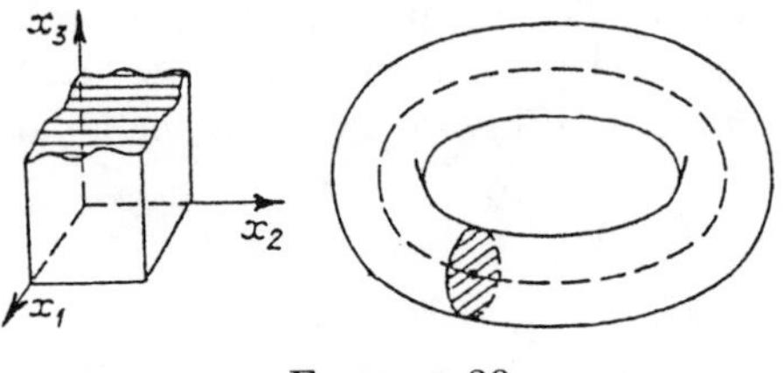

FIGURE 30

EXAMPLE 57. Let G be a Schottky group of genus $p \geq 1$ obtained from the cyclic loxodromic groups $\{g_1\}, \ldots, \{g_p\}$ by free combination. Then (the isometric spheres) $I(g_1), I(g_1^{-1}), \ldots, I(g_p), I(g_p^{-1})$ bound a fundamental polyhedron $P(G) \subset \mathbf{R}^3_+$ of it.[13] Identifying the faces of this polyhedron, we obtain a manifold $M(G)$ homeomorphic to $D_p \times [0, 1]$, where D_p is a closed disk with p open disks removed. This manifold is called a handlebody of genus p (see Figure 31). As above, it becomes a hyperbolic manifold by removing the boundary $\partial M(G)$, which is homeomorphic to a compact Riemann surface of genus p.

[13] *Translation editor's note*: It is assumed that $I(g_1), \ldots, I(g_p^{-1})$ are all disjoint.

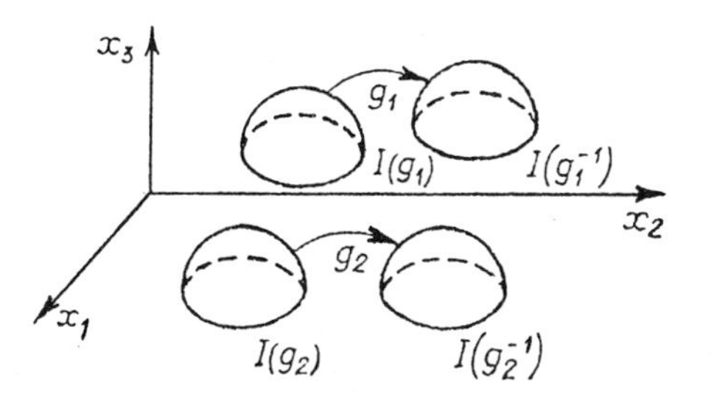

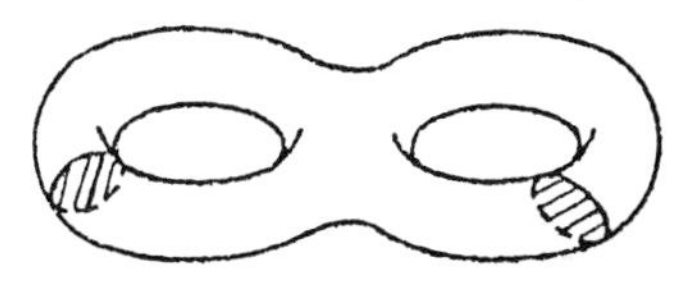

FIGURE 31

EXAMPLE 58. *A hyperbolic dodecahedron space.* By identifying sides of a dodecahedron we can obtain a manifold of constant negative curvature whose universal covering is a hyperbolic space. To do this consider the unit ball B^3 with the hyperbolic metric introduced there. In B^3 take a regular hyperbolic dodecahedron with center at the origin. It is completely determined by its hyperbolic diameter d. If we let d go to zero, then the hyperbolic dodecahedron tends to a Euclidean dodecahedron (the dihedral angles $\approx 117°$); but if d tends to infinity, then the dihedral angles of the dodecahedron tend to $2\pi/6$—the dihedral angles of a dodecahedron with vertices on the boundary of B^3. Thus, we can choose a value of d such that the dihedral angles are equal to $2\pi/5$.

For each two opposite faces of the dodecahedron, for example A and A^{-1}, we consider the loxodromic mappings $g_A = U \circ O \circ J$, where $J(x)$ is inversion with respect to the sphere (hyperbolic plane) containing A, $O(x)$ is reflection with respect to the plane perpendicular to the straight line joining the centers of A and A^{-1}, and $U(x)$ is rotation about this line through an angle equal to $3\pi/5$. Figure 32 shows the dodecahedron net, which completely determines the dodecahedron space obtained when faces are identified by the mappings g_A. It is clear that the edges of the dodecahedron fall into six classes consisting of five equivalent edges, and all the vertices are mutually equivalent. From this it is clear that our dodecahedron with the specified identification of faces is a fundamental region of the group $G \subset \mathcal{M}_3$ generated by the mappings g_A, which is therefore a Kleinian group in the ball B^3. Thus, the dodecahedron space we have constructed is a compact closed manifold of constant negative curvature whose universal covering space is the ball B^3 (hyperbolic space). It can be shown that its one-dimensional homology group is isomorphic to the direct sum of three cyclic groups of order 5 (see Exercise 130 and Problem 85).

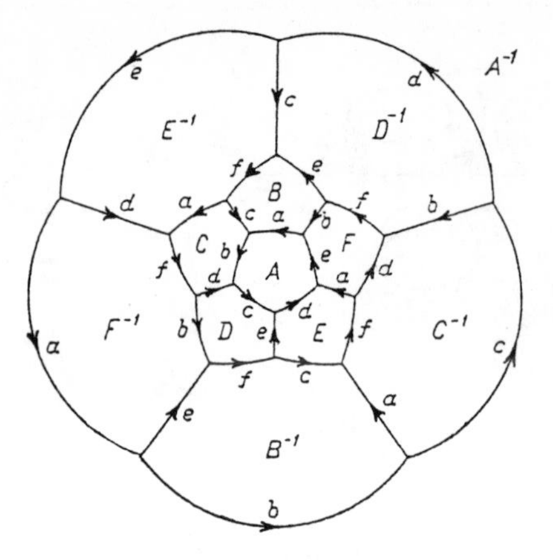

FIGURE 32

EXAMPLE 59 (RILEY [211] AND WEILENBERG [246]). *The Whitehead link.* Consider the subgroup G of the Picard group (see Example 15) of finite index generated by the Möbius mappings corresponding to the matrices

$$g_1 = \begin{pmatrix} 1 & 2 \\ 0 & 1 \end{pmatrix}, \quad g_2 = \begin{pmatrix} 1 & i \\ 0 & 1 \end{pmatrix}, \quad g_3 = \begin{pmatrix} 1 & 0 \\ -i-1 & 1 \end{pmatrix}.$$

The group G has the relations

$$g_3(g_1g_2^{-1})g_3g_2^{-1}g_3^{-1}g_2g_3^{-1}g_2^{-1} = 1, \qquad g_1g_2g_1^{-1}g_2^{-1} = 1. \tag{77}$$

Its fundamental polyhedron $F(G) \subset \mathbf{R}^3_+$ can be represented as the intersection of the exteriors of the isometric spheres $I(g_3)$ and $I(g_3^{-1})$ of q_3 with the polyhedra $\{x \in \mathbf{R}^3_+ : -1 < x_1 < 0, 0 < x_2 < 1\}$ and $\{x \in \mathbf{R}^3_+ : 0 < x_1 < 1, -1 < x_2 < 0\}$. This can be seen with the help of the Poincaré theorem. The projection of $F(G)$ on the boundary plane $\overline{\mathbf{C}} = \partial\mathbf{R}^3_+$ is shown in Figure 33. Here the numbers indicate the succession of identifications of the edges of $F(G)$ on $I(g_3)$ and $I(g_3^{-1})$ that corresponds to the first of the relations (77). The presentation (77) of G can be transformed to the form

$$G = \{g_2, g_3 \colon (g_2^{-1}g_3g_2g_3^{-1})(g_2g_3g_2^{-1}g_3^{-1})(g_2g_3^{-1}g_2^{-1}g_3)(g_2^{-1}g_3^{-1}g_2g_3)\}. \tag{78}$$

From this it is clear that G is anti-isomorphic to the fundamental group $\pi_1(S^2\backslash l)$ of the Whitehead link, which has the presentation

$$\{a_1, a_2 \colon (a_2^{-1}a_1a_2a_1^{-1})(a_2^{-1}a_1^{-1}a_2a_1)(a_2a_1^{-1}a_2^{-1}a_1)(a_2a_1a_2^{-1}a_1^{-1})\}, \tag{79}$$

which is analogous to (78). The indicated anti-isomorphism is obtained from the correspondence $g_3 \leftrightarrow a_2^{-1}$ and $g_2 \leftrightarrow a_1$ with (78) and (79) taken into account. From this and from Riley's theorem [211] (see Problem 86) it follows that of the manifold $M(G) = \mathbf{R}^3_+/G$ is homeomorphic to the exterior of the Whitehead link (see Figure 34).

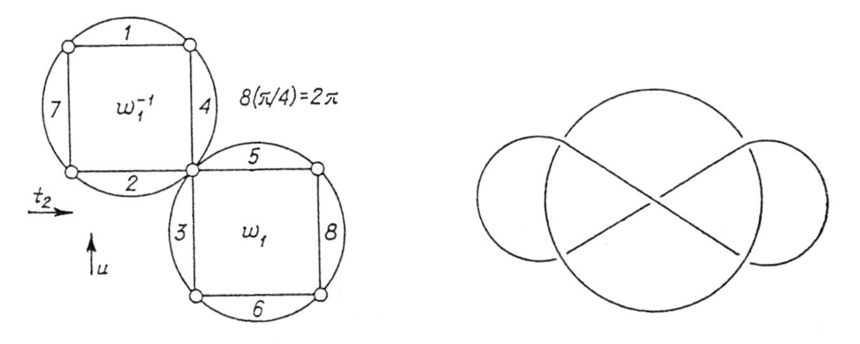

FIGURE 33 FIGURE 34

EXAMPLE 60 (RILEY [211] AND WIELENBERG [246]). *The Borromean rings.* Consider the subgroup G of the Picard group (see Example 15) of finite index that is generated by the Möbius mappings given by the matrices

$$t_4 = t^4 = \begin{pmatrix} 1 & 4 \\ 0 & 1 \end{pmatrix}, \quad u_2 = u^2 = \begin{pmatrix} 1 & 2i \\ 0 & 1 \end{pmatrix},$$

$$v = ata^{-1} = \begin{pmatrix} 1 & 0 \\ -1 & 1 \end{pmatrix}, \quad s = (t^{-1}u^{-1})v(ut) = \begin{pmatrix} 2+i & 2i \\ -1 & -i \end{pmatrix},$$

and has the presentation

$$G = \{t_4, u_2, v, s\colon t_4u_2 = u_2t_4,\ svs^{-1} = vt_4,\\ s(u_2^{-1}vu_2)s^{-1} = u_2^{-1}vt_4u_2\}. \tag{80}$$

To represent its structure and also that of the manifold $M(G) = \mathbf{R}^3_+/G$ more clearly we consider the subgroup $G_1 \subset G$ having presentation $G_1 = \{t_4, u_2, v\colon t_4u_2 = u_2t_4\}$. The intersection of the exteriors of the isometric spheres $I(v)$ and $I(v^{-1})$ (their projections on the plane $\mathbf{C}$ are $\{|z-1| = 1\}$ and $\{|z+1| = 1\}$) with the prism

$$P_\infty = [x \in \mathbf{R}^3_+\colon\ -2 < x_1 < 2, -1 < x_2 < 1\}$$

can be taken as the fundamental polyhedron $F(G_1)$ (see Figure 35). Then $M(G_1)$ has two components of the boundary $\partial M(G_1)$, both homeomorphic to a sphere with three points deleted. To these components there correspond the nonconjugate free Fuchsian subgroups Γ_1 and Γ_2 generated by $v, u_2^{-1}vu_2$ and $vt_4, u_2^{-1}(vt_4)u_2$, respectively, and leaving invariant the circles $|z+i| = 1$ and $|z+2+i| = 1$, respectively.

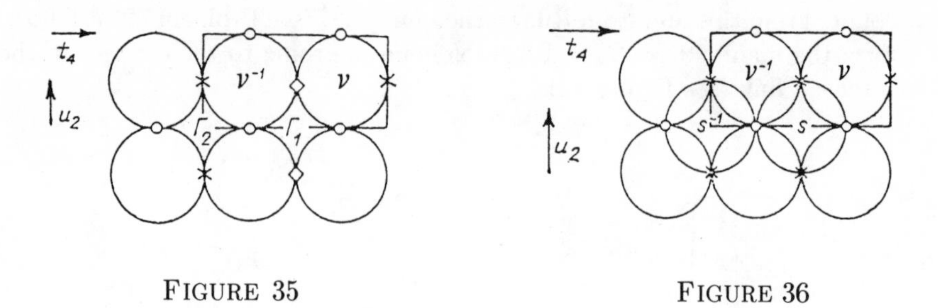

FIGURE 35 FIGURE 36

We now observe that G is the extension of the subgroup G_1 by an element s that conjugates the Fuchsian subgroups $\Gamma_1, \Gamma_2 \subset G_1 : \Gamma_2 = s\Gamma_1 s^{-1}$. The discrete group G no longer acts discontinuously in $\overline{\mathbf{C}}$, and $M(G)$ is the manifold without boundary obtained from $M(G_1)$ by gluing the two components of $\partial M(G_1)$ together. The intersection of the exteriors of the isometric spheres $I(v), I(v^{-1}), I(s^{-1}) = \{|z+2+i| = 1\}$, and $I(s) = \{|z+i| = 1\}$, and their images under the translations u_2 and t_4 with the prism P_∞, can be taken as the fundamental polyhedron $F(G)$ (Figure 36). But from this we get that the presentation (80) can be rewritten in the form

$$G = \{t_4, u_2, v, s : t_4 = v^{-1}svs^{-1}, v(u_2 s u_2^{-1} s^{-1}) = (u_2 s u_2^{-1} s^{-1})v, \\ (v^{-1}svs^{-1})u_2 = u_2(v^{-1}svs^{-1})\}. \quad (81)$$

([14]) The presentation of the fundamental group $\pi_1(S^3 \backslash l)$ of the Borromean rings can be written in the form

$$\{x_1, x_2, x_3 : x_1 = (x_2^{-1}x_3x_2x_3^{-1})x_1(x_3x_2^{-1}x_3^{-1}x_2), \\ x_2 = (x_3^{-1}x_1^{-1}x_3x_1)x_2(x_1^{-1}x_3^{-1}x_1x_3)\}. \quad (82)$$

If we now consider the correspondence $u_2 \to x_1, v \to x_2^{-1}, s \to x_3^{-1}$, then with the presentations (81) and (82) taken into account it establishes an anti-isomorphism between G and $\pi_1(S^3 \backslash l)$. From this and Riley's theorem [211] (see Problem 86) it follows that $M(G) = \mathbf{R}^3_+/G$ is homeomorphic to the exterior of the Borromean rings (see Figure 37).

([14]) *Translation editor's note*: It appears that the second line of (81) should read $v(s^{-1}u_2su_2^{-1}) = (s^{-1}u_2su_2^{-1})v$.

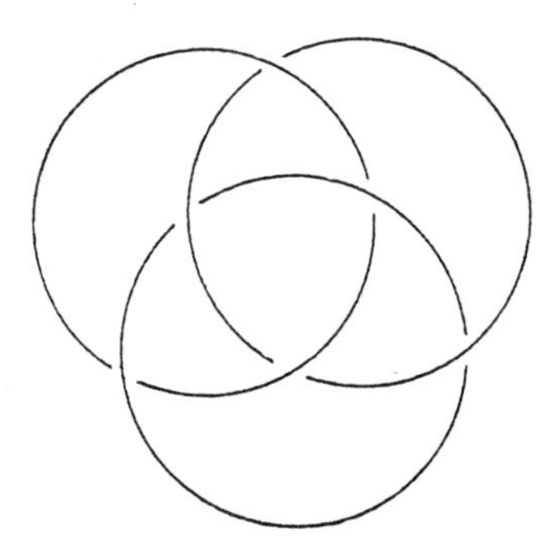

FIGURE 37

EXAMPLE 61 (WIELENBERG [246]). *A closed chain of four links.* Suppose that the group $G \subset \mathcal{M}_3$ is generated by the mappings that correspond in the group $SL(2, \mathbf{C})$ to the matrices

$$t_2 = \begin{pmatrix} 1 & 2 \\ 0 & 1 \end{pmatrix}, \qquad u_2 = \begin{pmatrix} 1 & 2i \\ 0 & 1 \end{pmatrix},$$
$$v_1 = \begin{pmatrix} 1 & 0 \\ -1+i & 1 \end{pmatrix}, \qquad w_1 = \begin{pmatrix} 1 & 0 \\ -1-i & 1 \end{pmatrix}.$$

The group G is a subgroup of the Picard group of finite index, because $v_1 = ata^2u^{-1}a$ and $w_1 = atua^{-1}$ (here $a(z) = -1/z, t(z) = z+1, u(z) = z+i$ and $l(z) = -z$ are the generators of the Picard group; see Example 15), and the fundamental polyhedron $F(G)$ of G equal to the intersection of the exteriors of the isometric spheres $I(v_1), I(v_1^{-1}), I(w_1), I(w_1^{-1})$ with the infinite prism $\{x \in \mathbf{R}^3_+ : |x_1| < 1, |x_2| < 1\}$ has finite hyperbolic volume (see Figure 38). Identification of faces in $F(G)$ yields the following relations in G:

$$t_2u_2t_2^{-1}u_2^{-1} = v_1w_1v_1^{-1}w_1^{-1} = u_2v_1t_2w_1u_2^{-1}v_1^{-1}t_2^{-1}w_1^{-1} = 1. \tag{83}$$

Writing these relations in the form

$$t_2u_2 = u_2t_2, \quad v_1w_1v_1^{-1} = w_1, \quad v_1(t_2w_1u_2^{-1})v_1^{-1} = u_2w_1t_2, \tag{84}$$

we get that G is the extension of the group

$$G_1 = \{t_2, u_2, w_1 : u_2(u_2^{-1}w_1t_2)t_2^{-1}w_1^{-1} = t_2^{-1}(t_2w_1u_2^{-1})u_2w_1^{-1} = t_2u_2t_2^{-1}u_2^{-1} = 1\} \tag{85}$$

(the projection of its fundamental polyhedron $F(G_1)$ is shown in Figure 39) by means of a mapping v_1 conjugating the nonequivalent free Fuchsian subgroups Γ_1 and Γ_2 of the group (85) with generators $w_1, t_2w_1u_2^{-1}$ and $w_1, u_2w_1t_2$, respectively.

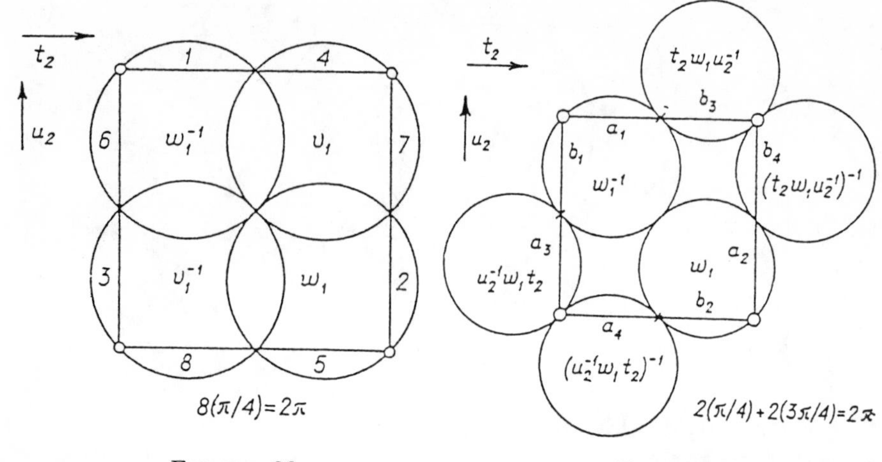

FIGURE 38 FIGURE 39

The manifold $M(G_1) = \mathbf{R}^3_+/G_1$ has two boundary components (corresponding to the Fuchsian stabilizers of Γ_1 and Γ_2 and homeomorphic to a sphere with three points deleted), and the manifold $M(G)$ is obtained from $M(G_1)$ by gluing these components together. The resulting 3-manifold $M(G)$ is homeomorphic to the exterior of the link pictured in Figure 40, which has fundamental group

$$\begin{aligned}\{a_1, a_2, a_3, a_4 \colon a_3a_4 = a_4a_3, a_1^{-1}a_2a_1 = a_2,\\ a_1^{-1}(a_4^{-1}a_2a_3)a_1 = a_3a_2a_4^{-1}\}.\end{aligned} \tag{86}$$

This follows from a theorem of Riley [211] (see Problem 86), a comparison of the group presentations (84) and (86), and the fact that these groups are anti-isomorphic.

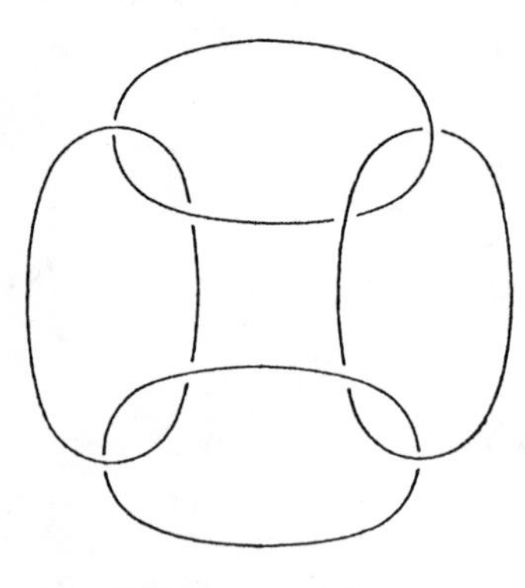

FIGURE 40

FIGURE 41

EXAMPLE 62 (WALDINGER [241] AND WIELENBERG [246]). *A closed chain of six links.* In the preceding example we obtained a uniformizatio by a Kleinian group of the complement of a closed chain made up of four links. We now determine a subgroup of the Picard group (of finite index) that uniformizes the exterior of the link shown schematically in Figure 41. Its fundamental group π_1 has the form

$$\pi_1 = \{a, b, c, d\colon ab = ba, cd = dc\}. \tag{87}$$

We first consider the subgroup G_1 of the Picard group with generators given by the matrices

$$t_2 = \begin{pmatrix} 1 & 1 \\ 0 & 1 \end{pmatrix}, \quad u_2 = \begin{pmatrix} 1 & i \\ 0 & 1 \end{pmatrix}, \quad v_2 = \begin{pmatrix} 1 & 0 \\ 2 & 1 \end{pmatrix}, \quad w_2 = \begin{pmatrix} 1 & 0 \\ -2i & 1 \end{pmatrix},$$

for which the commutators $[t_2, u_2]$ and $[v_2, w_2]$ are equal to I. The boundary $\partial M(G_1)$ of the manifold $M(G_1) = (\mathbf{R}^3_+ \cup \Omega(G_1))/G_1$ has one component $S(G_1)$, which is homeomorphic to a sphere with four points deleted. The presentation of the corresponding Fuchsian stabilizer Γ can be written in the form $\Gamma = \{\gamma_1, \gamma_2, \gamma_3, \gamma_4\colon \gamma_4\gamma_3\gamma_2\gamma_1 = 1\}$. The group Γ leaves invariant the circle $\{z\colon |z + i - 1| = 1\}$, and the mappings $\gamma_1 = t_2 v_2$, $\gamma_2 = t_2(u_2^{-1} w_2^{-1}) t_2^{-1}$, $\gamma_3 = u_2^{-1}(v_2^{-1} t_2^{-1}) u_2$, and $\gamma_4 = w_2 u_2$ are generators of it (see the scheme in Figure 42).

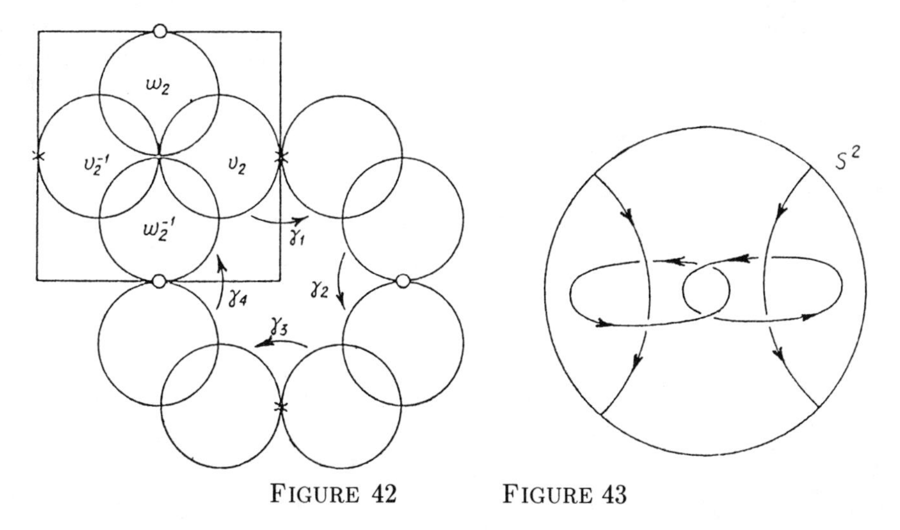

FIGURE 42 FIGURE 43

Now let J be inversion with respect to the circle $\{|z + i - 1| = 1\}$, and let $G_2 = J G_1 J^{-1}$. Note that under this conjugation the elements of Γ are carried into themselves, and G_2 is again a subgroup of the Picard group, because $G_2 = (tu^{-1}aut^{-1}) G_1 (tu^{-1}aut^{-1})$ (here a, u, and t are the same as in the preceding example). Consider now the free product G of the groups G_1 and G_2 with amalgamated Fuchsian subgroup Γ, i.e., the discrete group G obtained from G_1 and G_2 by Maskit combination with amalgamated subgroup

Γ. It can be shown [241] that the index of G in the Picard group is equal to 48.

Since $M(G_1)$ has two toroidal punctures and two cylindrical ones (see Figure 43 and cf. Examples 54 and 56), and its boundary $\partial M(G_1)$ is an incompressible surface, the manifold $M(G)$ corresponding to G has fundamental group (87), i.e., the complement of the link shown in Figure 41.

EXAMPLE 63 (JØRGENSEN [110]). *A compact hyperbolic manifold which is fibered over a circle.* The indicated manifold is constructed as the space of orbits of the discrete group G^* of isometries of hyperbolic space that was constructed in Example 53 (as a cyclic extension of G that is not geometrically finite). The intersection of the exteriors of the isometric spheres of the mappings in the set (74) with the dihedral angle whose faces are carried one into the other by the rotation K serves as the fundamental polyhedron of G (see Figure 27); the intersection of the polyhedron Q with the shell between hemispheres about zero carried one into the other by the transformation $T(x) = \rho^2 x$ can be taken as the fundamental polyhedron of G^*. The boundary of Q consists of hexagons having axes of symmetry. These axes are fixed lines for second-order elliptic transformations obtained with the help of the Lie product of the generators A and B by the formula $\varphi = AB - BA, \varphi^2 = I$. Considering on the boundary of Q any hyperbolic polygonal line perpendicular to the corresponding lines of symmetry of the boundary hexagons, we get a fibering of the part of the boundary of Q under consideration into polygonal arcs (see Figure 27). This fibering is not hard to extend to Q itself. By considering the identification of boundary points of the resulting fibers by transformations in G (see the schemes in Figure 44), it is easy to see that each such fiber is a torus with one branch point of order m, and hence $\mathbf{R}^3_+/G$ is a fibering over the real line whose fibers are tori with one branch point of order

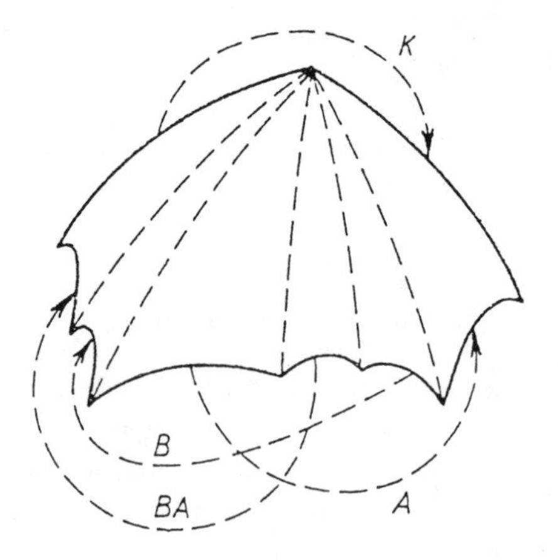

FIGURE 44

m. Taking into account the periodicity of this fibering (which corresponds to the action of T), we get that the compact manifold $\mathbf{R}^3_+/G^*$ can be represented as a fibering over the circle with fibers that are tori with one branch point.

The manifold thus constructed is a branched manifold. To construct an unbranched (smooth) manifold that is fibered over the circle, we make the following changes in the construction of the groups.

It follows from a theorem of Selberg [219] that there exists a torsion-free normal subgroup $N^* \subset G^*$ of finite index. The subgroup $N = N^* \cap G \subset G$ is thus also normal, and $[G : N] < \infty$. Denote by $N(C)$ the group $\langle N, C \rangle$, where $C = T^k \in N^* \backslash \{I\}$ for some integer k. Then N is a normal subgroup of $N(C)$, and $N(C)/N$ is an infinite cyclic group. Since the groups $N(C)$ and N are torsion-free,

$$\pi_1(\mathbf{R}^3_+/N(C)) \simeq N(C), \qquad \pi_1(\mathbf{R}^3_+/N) \simeq N. \tag{88}$$

Moreover, $\mathbf{R}^3_+/N(C)$ is a smooth compact manifold, and it follows from a theorem of Stallings on fibering and from (88) that there exists a compact surface S such that $N \simeq \pi_1(S)$, while $\mathbf{R}^3_+/N(C)$ can be represented as a fibering over the circle with fibers homeomorphic to S.

These manifolds $\mathbf{R}^3_+/G^*$ and $\mathbf{R}^3_+/N(C)$ are the first examples of compact hyperbolic manifolds fibered over the circle.

EXAMPLE 64 (THURSTON [234]). *A hyperbolic structure on the complement of the figure-eight knot.* We take two tetrahedra and glue together their faces as shown in Figure 45. The result is a complex consisting of two tetrahedra, four triangles, two edges, and one vertex. It is not a manifold; its Euler characteristic is equal to 1, and a neighborhood of the vertex is a cone over a torus. Removal of the vertex turns this complex into a manifold M which represents the complement of a figure-eight knot (see Figure 46). This becomes clearer if we see how this knot is stretched onto a tetrahedron with faces A, B, C, D. Figure 47 shows the form of this stretching from above. It is obtained by joining the links of the knot by means of oriented segments, and stretching four membranes A, B, C, and D onto the resulting one-dimensional complex. The boundary of each membrane consists of the indicated segments, joined by the parts of the knot corresponding to the removed vertices of the tetrahedron.

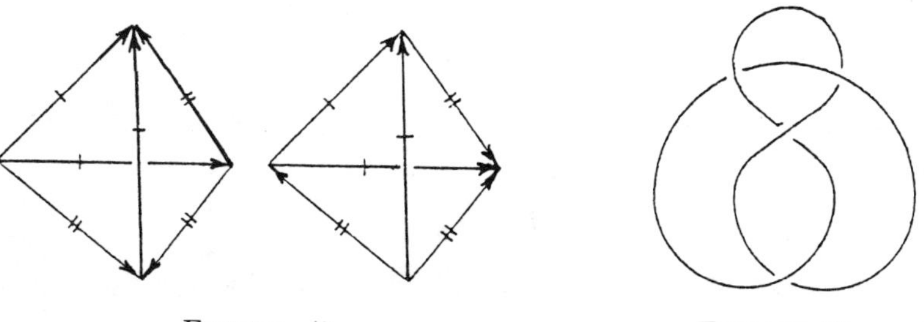

FIGURE 45 FIGURE 46

Since six dihedral angles meet at each edge of the complex, tetrahedra with dihedral angles equal to $\pi/3$ are needed for a conformal gluing in this way. Such tetrahedra can be taken to be regular hyperbolic tetrahedra with vertices on the sphere at infinity (see Figure 48). It is easy to construct hyperbolic motions realizing the needed gluing. The group G they generate is Kleinian (by the Poincaré theorem). The manifold $M = B^3/G$ thus has a hyperbolic structure and is uniformized by the Kleinian group G.

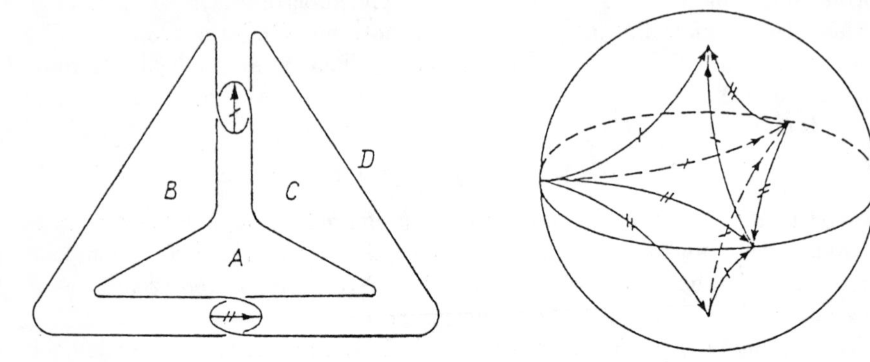

FIGURE 47

FIGURE 48

EXAMPLE 65 (THURSTON [234]). *A manifold with hyperbolic structure having a geodesic boundary.* There are many ways similar to the way in the preceding example for gluing boundaries of tetrahedra. We can determine a gluing (see Figure 49) such that the resulting complex consists of two tetrahedra, four triangles, one edge, and one vertex (its Euler characteristic is equal to 2). It turns out that removal of a certain neighborhood of the vertex turns this complex into a manifold M having geodesic boundary, i.e., the double of this manifold has a hyperbolic structure in which the gluing surface in the double is geodesic. This corresponds to the fact that a discrete group G of hyperbolic motions (i.e., a Fuchsian group in the ball B^3) that uniformizes the manifold M ($M = B^3/G$) has the following properties:

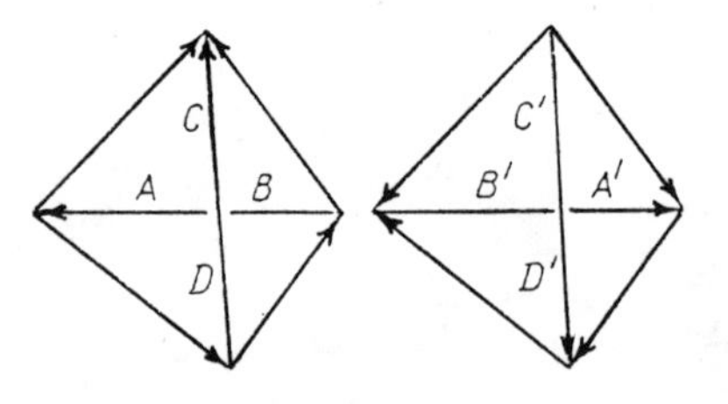

FIGURE 49

a) The restriction of G to the boundary ∂B^3 of the ball is a Kleinian group $\tilde{G}$ on the sphere.

b) $\Omega(\tilde{G})$ has a single class $\{\Omega_0\}$ of equivalent components.

c) The stabilizer of the component Ω_0 in $\tilde{G}$ is a Fuchsian group on the sphere that uniformizes the indicated gluing surface.

EXAMPLE 66 (THURSTON [234]). *An octahedron hyperbolic space that is homeomorphic to the complement of the Whitehead link.* Suppose that we have an octahedron whose faces are identified in the way indicated on the octahedron development shown in Figure 50. Here the point at infinity is one of the vertices. Then we obtain a complex consisting of one octahedron, four triangles, three edges, and two vertices; its characteristic is equal to 2. Removal of the vertices turns it into a manifold M on which a hyperbolic structure can be introduced. This follows from the fact that the indicated gluing can be performed on a regular octahedron with vertices lying on the unit sphere by considering that the octahedron is imbedded in the projective model of hyperbolic space (when the geodesics are chords) and, consequently, has all dihedral angles equal to $\pi/2$ (see Figure 51). Moreover, the gluing mappings (motions in this model of hyperbolic space) generate a discrete group G (the Poincaré theorem), and $M = B^3/G$. Stretching the Whitehead link onto the octahedron as in Example 64 (see Figure 52), we can show that this manifold is homeomorphic to its complement. The indicated stretching of the link onto the octahedron can be realized as follows: Join the components of the link by oriented segments as shown in Figure 53, and stretch the four membranes A, B, C, and D onto the resulting one-dimensional complex; the boundaries of the membranes consist of the indicated oriented segments, joined by the link parts. These membranes correspond to the four triangles of our complex obtained by gluing faces of the octahedron, and the parts of their boundaries joining the oriented segments correspond to the two removed vertices of the complex. Thus, we have obtained on the complement of the Whitehead link (cf. Example 59) a hyperbolic structure induced by the action of a group of isometries on the projective model of hyperbolic space.

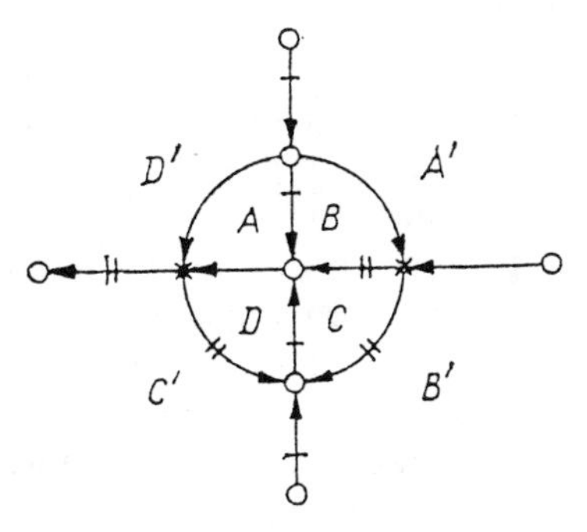

FIGURE 50

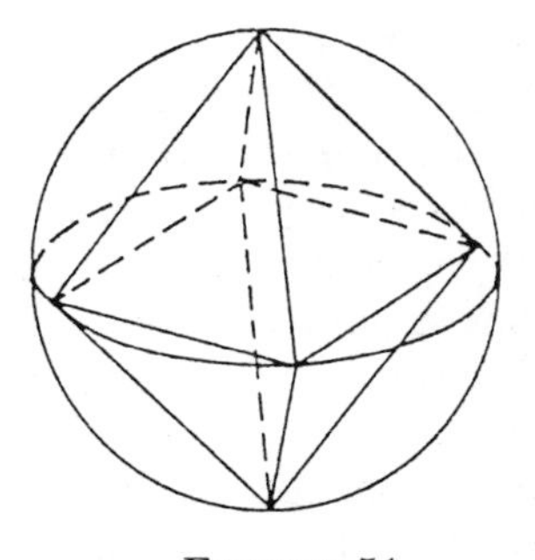

FIGURE 51

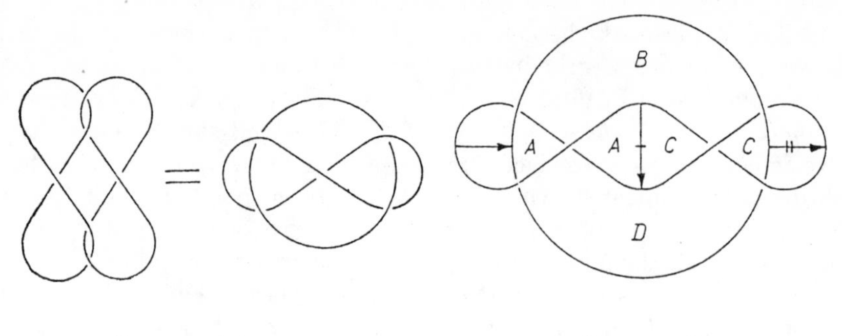

FIGURE 52

FIGURE 53

EXAMPLE 67 (THURSTON [234]). *A hyperbolic structure on the complement of the Borromean rings—a projective approach.* Take two octahedra and glue their corresponding faces after first rotating one of them through the angle $2\pi/3$. Such a correspondence of their faces is shown on a spherical development of the octahedra in Figure 54. The resulting complex (two octahedra, eight triangles, six edges, and three vertices) has Euler characteristic 3, and removal of its three vertices turns it into a manifold M. As above, we join the link components by six oriented segments and as a result obtain the one-dimensional complex given in Figure 55. Here it is shown how two-dimensional cells must be stretched on this complex in order to get a development of the octahedron. Furthermore, the boundaries of the cells being stretched consist of the oriented segments joined by parts of the link circles. These parts correspond to the vertices on the development obtained. This makes it clear that a manifold M homeomorphic to the complement of the Borromean rings (this link is shown in Figure 37) is obtained by gluing together two three-dimensional cells whose boundaries are decomposed in the indicated way into two-dimensional cells. If our octahedra are regarded as imbedded in the projective model of hyperbolic space (as in the preceding example, the ball B^3 with a metric in which chords are geodesics and the boundary $\partial B^3 = S^2$ is the ideal) and as having ideal vertices and dihedral angles equal to $\pi/2$, then the indicated gluing can be carried out by motions of this model of hyperbolic space (by projective transformations of the ball). This operation introduces a complete hyperbolic structure on the complement of the Borromean rings, and the manifold M can be represented in the form $M = B^3/G$, where G is the discrete group of hyperbolic isometries generated by the gluing mappings. A complete hyperbolic structure was introduced earlier on the same manifold (in Example 60) with the help of a theorem of Riley which establishes a connection between the hyperbolic structure and an anti-isomorphic imbedding of the fundamental group $\pi_1(M)$.

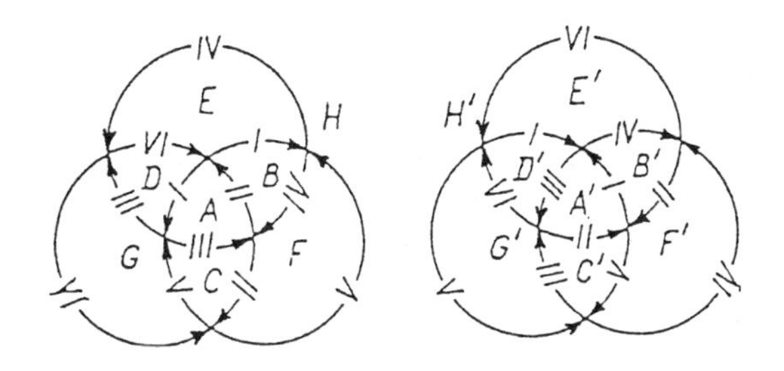

FIGURE 54

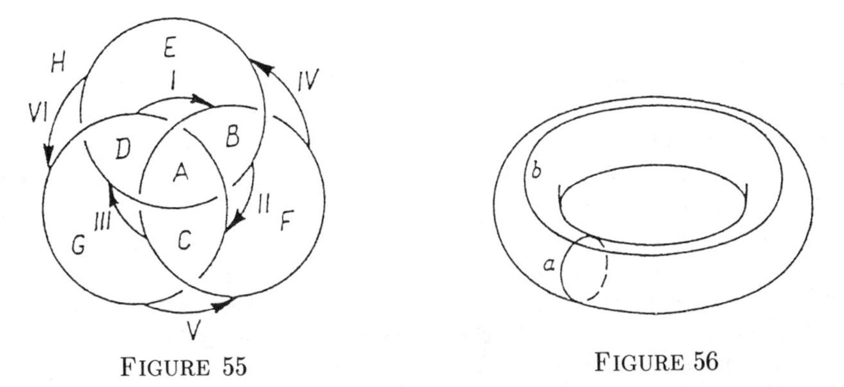

FIGURE 55

FIGURE 56

EXAMPLE 68. *Dehn surgery on the complement of the figure-eight knot.* Suppose that M is the complement of an m-component link in S^3, P_i is a tubular neighborhood of the ith component of the link, $T_i^2 = \partial P_i, a_i$ is a meridian ($a_i \simeq 0$ in P_i) on T_i^2 and $\bar{b}_i$ is a parallel, and they are taken as the generators of $H_1(T_i^2, \mathbf{R})$ (see Figure 56). Denote by $M_{(\alpha_1,\beta_1),\dots,(\alpha_m,\beta_m)}$ the manifold obtained by gluing solid tori to $M \setminus \bigcup_i (\operatorname{int} P_i)$ in such a way that the meridian of the ith torus being glued goes into $\alpha_i a_i + \beta_i b_i, (\alpha_i, \beta_i) \in \mathbf{R}^2$. But if the ith link component is not affected, then the pair $d_i = (\alpha_i, \beta_i)$ is denoted by ∞. Thus, $M = M_{\infty,\dots,\infty}$. This procedure is called *generalized Dehn surgery on* M, and the vector $(d_1, \dots, d_m)$ is *the parameter of the Dehn surgery*.

The question arises as to how many of the manifolds $M_{d_1,\dots,d_m}$ are hyperbolic if the original manifold M has a hyperbolic structure. Thurston [234] obtained the answer to this question for the complement of the figure-eight knot (see Example 64) by using foliation techniques. It turns out that the closed manifolds thereby obtained are irreducible and (except for six) are hyperbolic. The following parameters correspond to the six that are not hyperbolic: a) $(1,0)$, the trivial surgery, $M_{(1,0)} = S^3$; b) $(1,1), (2,1), (3,1)$, surgeries giving Seifert fiberings, and two sufficiently large manifolds (see Problem 118); c) $(0,1)$, the fibering into tori over the circle; and d) $(4,1)$, a so-called graph manifold. We remark that $M_{(\alpha,\beta)}, (\alpha,\beta) \neq (0,1), (4,1)$, give examples of irreducible manifolds that are not sufficiently large and are not Seifert fiberings.

EXAMPLE 69. *Hyperbolic structures on compact manifolds close to being hyperbolic.* Suppose that $H = S^3 \backslash l$ is a noncompact manifold, where l is some m-component link, and suppose that this manifold admits a complete hyperbolic structure. We perform on M generalized Dehn surgery with the parameters indicated in the preceding example: $d_1 = (p_1, q_1), d_2 = (p_2, q_2), \dots, d_m = (p_m, q_m), d_i \in \mathbf{R}^2$. This yields a closed manifold denoted by $M_{d_1,\dots,d_m}$. If $(d_1, \dots, d_m)$ tends to $(\infty, \dots, \infty)$ in the topology of the Cartesian product $S^2 \times S^2 \times \cdots \times S^2$, then the manifolds $M_{d_1,\dots,d_m}$ become close to the manifold $M = M_{\infty,\dots,\infty}$. This closeness is underscored in the following result of Thurston [234]: *There exists a neighborhood $U \subset S^2 \times \cdots \times S^2$ of the point $(\infty, \dots, \infty)$ such that any closed manifold $M_{d_1,\dots,d_m}$, $(d_1, \dots, d_m) \in U$, admits a hyperbolic structure.* It is interesting to compare this result with the following example and with Example 88, which shows that infinitely many manifolds having nonhyperbolic structure can be obtained from a hyperbolic manifold by generalized Dehn surgery.

EXAMPLE 70 (THURSTON [234]). *Complements of knots obtained by Dehn surgery from the complement of the Whitehead link.* As already shown in Examples 59 and 66, the complement of the Whitehead link (see Figures 34 and 52) admits a complete hyperbolic structure. It is clear from the preceding example that manifolds M_{d_1,d_2}, $(d_1, d_2) \in U(\infty, \infty)$, close to it are also hyperbolic. For $d_2 \in U(\infty) \in S^2$ the noncompact manifolds M_{∞,d_2}, which are the complements to the knots shown schematically in Figure 57 (the first is the figure-eight knot), also admit a complete hyperbolic structure. If d_2 tends to ∞, then $M_{\infty,d}$ tends to the manifold $M = M_{\infty,\infty}$ that is the complement of the Whitehead link. In fact, a stronger assertion holds. Namely, by computations using foliation techniques it can be shown in a way analogous to that in Example 68 that all these manifolds have a complete hyperbolic structure.

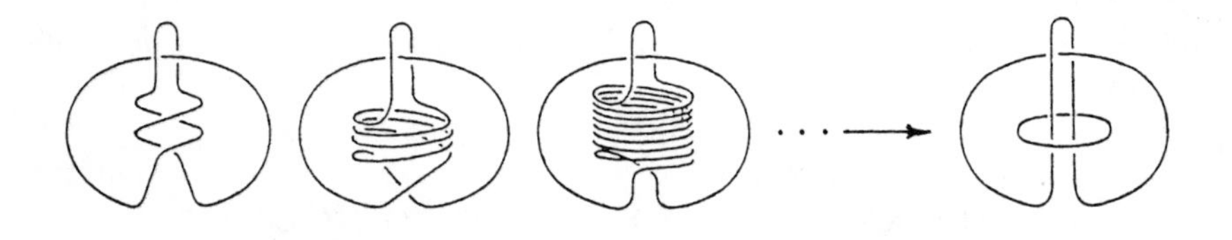

FIGURE 57

EXAMPLE 71 (GUSEVSKIĬ[91]). *The structure of a manifold that is uniformizable by a degenerate group.* The following is a well-known problem (see, for example, [59]): What is the structure of the manifold $M(G) = (B^3 \cup \Omega)/G$ for a degenerate group G and, in particular, what is the character of noncompactness of $M(G)$? It is shown in [91] that if $S = \Omega/G$ is a closed surface, then $M(G)$ is homeomorphic to $S \times [0,1)$; this corresponds to the case of a degenerate group without parabolic elements.

To prove this we construct (by means of cutting and gluing) a decreasing sequence M_j ($j = 0, 1, \ldots$) of neighborhoods of an end of $M(G)$ with the following properties:

1) M_j is a three-dimensional polyhedral manifold with compact connected boundary ∂M_j, $M(G) = M_0 \supset M_1 \supset M_2 \supset \cdots$, $\bigcap_0^\infty M_j = \varnothing$, and $\partial M_i \cap \partial M_j = \varnothing$ for $i \neq j$.

2) The homomorphism $\pi_1(\partial M_j) \to \pi_1(M(G))$ induced by the inclusion $\partial M_j \subset M(G)$ is a monomorphism for all $j = 0, 1, \ldots$.

By using some homotopy conditions it can now be shown that the manifold $M(G)$ can be "added together" from the cylinders $K_i = M_i \backslash \operatorname{int} M_{i+1}$ with bases homeomorphic to S, and, consequently, $M(G)$ is homeomorphic to $S \times [0, \infty)$.

We remark that removal of the boundary $\partial M(G)$ of $M(G)$ yields a hyperbolic manifold homeomorphic to a manifold $M(G_1) \backslash \partial M(G_1)$ that is uniformizable by a Fuchsian group G_1.

A similar result on the structure of the ends of the three-dimensional hyperbolic manifold $\mathbf{R}^3_+/G$ (that they have neighborhoods homeomorphic to the product of some surface by the interval $[0, \infty)$) was proved by Thurston [234] in the case of discrete groups $G \subset \mathcal{M}_2$ lying in the closure $T(\Gamma)$ of the Teichmüller space of any geometrically finite Kleinian group $\Gamma \subset \mathbf{M}_2$.

§6. Uniformization: Elliptic manifolds

In this section we give some examples of manifolds admitting an elliptic structure, i.e., having a sphere of appropriate dimension as universal covering space and having spherical isometries as covering transformations.

EXAMPLE 72. *Lens spaces.* Let $S^3 = \{(z_0, z_1) \in \mathbf{C}^2 \colon |z_0|^2 + |z_1|^2 = 1\}$ be the three-dimensional sphere, and let p and q be relatively prime integers. Define a mapping $h \colon S^3 \to S^3$ by

$$h(z_0, z_1) = (e^{2\pi i/p} z_0, e^{2\pi i q/p} z_1). \tag{89}$$

Then h is an automorphism of S^3 with period p (i.e., $h^p = 1$), and the group $\mathbf{Z}_p$ acts on S^3 without fixed points. The space of orbits of the action of $\mathbf{Z}_p$ on S^3 defined by (89) is called the *lens space* $L(p, q)$.

This three-dimensional elliptic manifold can also be defined in another way. To do this, consider a lens in $\mathbf{R}^3$, i.e., a region bounded by two spherical segments, and break up the rim of the lens into p equal arcs, thus turning these segments into two p-gons, A and A'. Reflecting A with respect to the plane of the rim of the lens and turning it through the angle $2\pi q/p$, we identify the sides of the lens. The manifold thus obtained is the lens space $L(p, q)$. If for $p = 2$ and $q = 1$ the lens is taken to be a ball, and its edge is an equatorial circle, then $L(2,1)$ is obtained as the projective plane $\mathbf{P}^3$ (the ball with h diametrically opposite points of the boundary sphere identified).

It is easy to see that $L(p,q)$ can be uniformized by a Kleinian group in $\mathbf{R}^3$. To do this it suffices to take a lens with dihedral angle at its rim equal to $2\pi/p$. The indicated identification is then realized by a Möbius mapping $g = U \circ O \circ J$, where $J(x)$ is inversion with respect to the sphere containing $A, O(x)$ is reflection with respect to the plane of the edge of the lens, and $U(x)$ is rotation about the axis of the lens through the angle $2\pi q/p$.

The cyclic group $G = \langle g \rangle \subset \mathcal{M}_3$ has order p, acts on S^3 without fixed points, and a lens is a fundamental region of it. Thus, the manifold S^3/G is homeomorphic to $L(p,q)$.

It is not hard to prove that the condition $q' = \pm q^{\pm 1}$ suffices for the lenses $L(p,q)$ and $L(p,q')$ to be homomorphic. Reidemeister showed that this condition is also necessary.

The following result of Whitehead provides a homotopy classification of lenses:

$L(p,q)$ is homotopy equivalent to $L(p,q')$ if and only if $q' = \pm m^2 q$ for some integer m.

In particular, this gives us that $L(5,1)$ and $L(5,2)$ have different homotopy types. We remark also that $L(7,1)$ and $L(7,2)$ give an example of nonhomeomorphic three-dimensional manifolds of the same homotopy type.

EXAMPLE 73. *Generalized lens spaces.* Let $S^{2n+1} = \{(z_0, z_1, \dots, z_n) \in \mathbf{C}^{n+1} \colon \sum |z_i|^2 = 1\}$ be the $(2n+1)$-dimensional sphere, and let $q_1, \dots, q_n$ be integers that are relatively prime to an integer p. Define a mapping $h \colon S^{2n+1} \to S^{2n+1}$ by

$$h(z_0, \dots, z_n) = (e^{2\pi i/p} z_0, e^{2\pi i q_1/p} z_1, \dots, e^{2\pi i q_n/p} z_n). \tag{90}$$

Here, as in the preceding example, the automorphism h of S^{2n+1} determines an action of the group $\mathbf{Z}_p$ on S^{2n+1} without fixed points. The orbit space $L(p, q_1, \dots, q_m)$ corresponding to the action (90) is called a *generalized lens space*. It can be shown [221] that its fundamental group is isomorphic to $\mathbf{Z}_p$.

EXAMPLE 74. *The projective space* $\mathbf{RP}^n$. Recall that the *projective space* $\mathbf{RP}^n$ is understood to be the space whose points are the straight lines in $\mathbf{R}^{n+1}$ through the origin O ($n \geq 2$); a neighborhood of a line l is understood to be the collection of all lines through O that intersect a neighborhood of some point on l different from O. The space $\mathbf{RP}^n$ can be obtained from the n-dimensional sphere S^n by identifying its diametrically opposite points, and also from the n-dimensional ball by identifying diametrically opposite points on its boundary S^{n-1}. In $\mathbf{RP}^n$ it is possible to introduce homogeneous coordinates by noting that the coordinates $x_0, x_1, \dots, x_n$ of different points on a line through the origin of $\mathbf{R}^{n+1}$ differ only by a common proportionality factor.

It is not hard to show that for even $n = 2m$ the projective space $\mathbf{RP}^n$ is nonorientable, while for odd $n = 2m+1$ it is an orientable manifold. Moreover, setting $p = 2$ and $q_1 = q_2 = \cdots = q_m = 1$ in the preceding example, we see that $\mathbf{RP}^{2m+1}$ can be obtained as the generalized lens space $L(2, 1, \ldots, 1)$.

But if we consider the Möbius mapping $g = O_1 \circ O_2 \circ \cdots \circ O_{2m-1} \circ J$, where $J(x) = x/|x|^2$ is inversion with respect to the sphere S^{2m} and $O_i(x)$ ($i = 1, \ldots, 2m+1$) is reflection with respect to the hyperplane through O and orthogonal to the ith coordinate axis, then it is clear that $\mathbf{RP}^{2m+1}$ is the space of orbits of the Kleinian group $G = \langle g \rangle$, which acts on S^{2m+1} without fixed points and has order 2; in other words, it can be uniformized by a Kleinian group. This gives us that the fundamental group $\pi_1(\mathbf{RP}^{2m+1})$ is isomorphic to $\mathbf{Z}_2$.

In the general case it is not hard to show that the homology groups $H_{2k+1}(\mathbf{RP}^n, \mathbf{Z})$ of odd dimension are cyclic groups of second order, while the homology groups $H_{2k}(\mathbf{RP}^n, \mathbf{Z})$ of even dimension are trivial. The zero-dimensional homology groups $H_0(\mathbf{RP}^n, \mathbf{Z})$ for any n and the n-dimensional groups $H_n(\mathbf{RP}^n, \mathbf{Z})$ for odd n are exceptions; these are always free cyclic groups.

As homology bases in odd dimensions we can take the coherently oriented projective subspaces $\mathbf{RP}^1, \mathbf{RP}^3, \mathbf{RP}^5, \ldots$.

EXAMPLE 75. *A Poincaré space with finite fundamental group.* This space is obtained from a dodecahedron by identifying opposite faces–pentagons rotated through an angle of $\pi/5$ with respect to each other. Under this identification (see Figure 58) there are 5 nonequivalent vertices, while the edges of the dodecahedron fall into 10 classes of three equivalent edges each. The Euler characteristic of this complex is equal to zero, so it is a manifold (see [217]), called a *spherical dodecahedron space.* This manifold was the first example of a Poincaré space, i.e., a manifold not homeomorphic to the sphere but having the same homology groups as the sphere: $H_0 = \mathbf{Z}$, $H_1 = H_2 = \mathbf{0}$, and $H_3 = \mathbf{Z}$. It was first obtained by Poincaré [205] in considering Heegaard diagrams in the form of a two-sheeted covering of the space of a spiral knot with invariants $(3, 5)$. Moreover, it can be shown that this manifold (dodecahedron space) is a branched p-sheeted covering of S^3 with branching along a torus knot with invariants (q, r), where p, q, r is any permutation of $2, 3, 5$. It was established in [218] that the manifolds thus obtained are homeomorphic to the dodecahedron space. Among the numerous examples of Poincaré spaces we now have, the spherical dodecahedron space is the only known manifold with finite fundamental group. Namely, it can be shown that it coincides with the icosahedron space which is the orbit space of the action of the binary icosahedron group on S^3. This group has order 120 and is characterized by the presentation $\{a, b, c\colon a^5 = b^2 = c^3 = abc\}$.

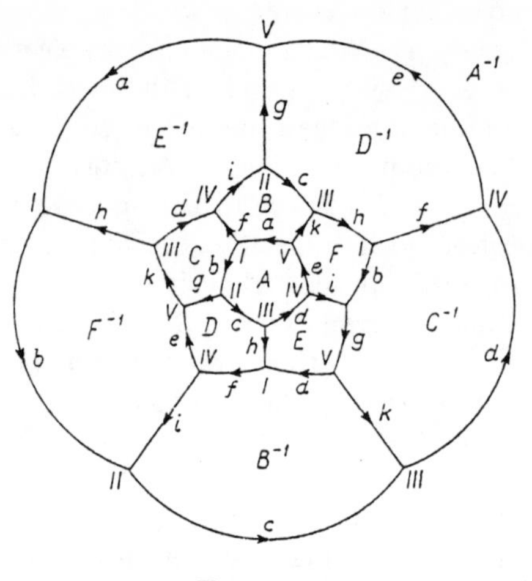

FIGURE 58

We show that a spherical dodecahedron space can be uniformized by a finite group $G \subset \mathcal{M}_3$. To do this it suffices to find a discontinuous group $G \subset \mathcal{M}_3$ whose generators realize the necessary identification of faces of the dodecahedron. Because of the triples of equivalent edges, the dihedral angles of the dodecahedron must equal $2\pi/3$. But since they are roughly equal to 117° for a Euclidean dodecahedron, it is clear that a spherical dodecahedron must be taken. We carry out the construction in the spherical space $\overline{\mathbf{R}}^3 = \mathbf{R}^3 \cup \{\infty\}$, in which the spherical planes are the spheres intersecting the sphere $x_1^2 + x_2^2 + x_3^2 = 1$ along its equators. The center of the dodecahedron is the origin of coordinates. If the Euclidean diameter of this spherical dodecahedron tends to zero, then its dihedral angles tends to 117° (the spherical planes tend to Euclidean planes); but if the diameter tends to 2, then our dodecahedron approximates the unit ball, and its dihedral angles approximate π. Thus, there is a value of the diameter of a spherical dodecahedron for which its dihedral angles are equal to $2\pi/3$. If for its equivalent faces, for example, A and A^{-1}, we now consider the mapping $g_A(x) = U \circ O \circ J(x)$, where $J(x)$ is inversion with respect to the sphere containing $A, O(x)$ is reflection with respect to the plane perpendicular to the line joining the centers of A and A^{-1}, and $U(x)$ is rotation about this line through the angle $\pi/5$, then the group $G \subset \mathcal{M}_3$ generated by these mappings is the desired group. This follows from the fact that the dodecahedron we have constructed is a convex fundamental polyhedron for it, and the sum of the dihedral angles of this polyhedron at equivalent edges is equal to 2π. Thus, the spherical dodecahedron space is represented in the form $\overline{\mathbf{R}}^3/G$, where $G \subset \mathcal{M}_3$ is a finite group isomorphic to the binary icosahedron group.

Finally, we note that a dodecahedron space (=an icosahedron space) can be given analytically as the intersection of the surface $z_1^2 + z_2^3 + z_3^5 = 0$ in $\mathbf{C}^3$ with the unit sphere $|z_1|^2 + |z_2|^2 + |z_3|^2 = 1$.

EXAMPLE 76. *A spherical octahedron space.* An octahedral space is obtained on identifying the opposite triangles of an octahedron, after rotation about the angle $\pi/3$ with respect to each other. Figure 59 shows the net of an octahedron obtained by stereographic projection on the plane; ∞ is also a vertex of the net, and the arrows show the orientation; the same letters denote equivalent edges (four classes with three edges each); all the vertices are equivalent here. From this it is clear that the dihedral angles of the octahedron must be equal to $2\pi/3$ (while they are approximately equal to $109°$ for a Euclidean octahedron). Thus, such an identification can be made only on a spherical octahedron. As in the preceding example, the finite group $G \subset \mathcal{M}_3$ generated by the four Möbius mappings $g_i = U_i \circ O_i \circ J_i$ identifying the faces of the octahedron in the indicated way is Kleinian, and the octahedral space can be represented as the manifold $\overline{\mathbf{R}}^3/G$, which has S^3 as universal covering.

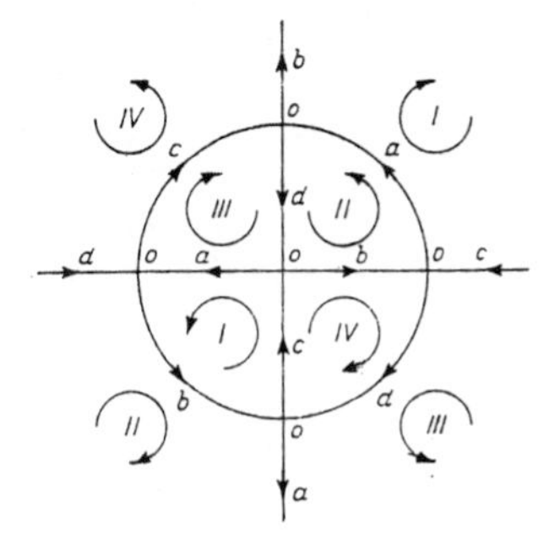

FIGURE 59

§7. General uniformization problems

EXAMPLE 77 (BELINSKIĬ [46]). *A universal Riemann surface containing arbitrary compact surfaces with boundaries as subregions.* In considering differential equations, especially nonlinear ones, on Riemann surfaces and in other problems it is necessary to work with variable surfaces running through some family of surfaces. Here it is more convenient to define the corresponding functions on a single surface containing all the surfaces in the given family as subregions. Below we construct a universal surface for all compact surfaces with boundaries; here surfaces are considered to within conformal equivalence.

A *surface of type* (g, n, m) is defined to be a Riemann surface obtained from some compact surface of genus g by cutting out n holes (simply connected regions with analytic boundaries) and deleting m points, where $g, n, m \geq 0$. It will be proved that the surface to be constructed has surfaces of type $(g, n, 0)$

as subregions, which suffices for our purposes because surfaces of type (g, n, m) are obtained by deleting additional points from surfaces of type $(g, n, 0)$; for simplicity (g, n) will be written instead of $(g, n, 0)$.

Let us first consider surfaces of a particular type (g, n) and single out a countable dense set $\{R_k\}$ of surfaces in the Teichmüller space T_g (of closed surfaces of genus g). Every surface $R(g, n)$ of type (g, n) with boundary can be extended to a closed surface $\hat{R}(g)$ of genus g (the holes are filled by gluing in disks). By varying the disks quasiconformally (only one of them would suffice), it is possible to vary the moduli of $\hat{R}(g)$ in such a way that the images of $\hat{R}(g)$ fill a whole neighborhood in the Teichmüller space (see [130]). In this neighborhood there is at least one surface in the countable basis $\{R_k\}$. In other words, all surfaces of type (g, n) are subregions of surfaces in the countable sequence $\{R_k\}$.

On each surface R_k take a countable dense set $\{p_{k,j}\}$, $j = 1, 2, \ldots$, of points with neighborhoods of hyperbolic radius $r_{k,j}$ such that $r_{k,j} \to 0$ as $j \to \infty$. Denote by $R_{k,j}$ the surface R_k with the (hyperbolic) disk about $p_{k,j}$ deleted. Since the centers of the disks are dense in R_k and their radii tend to zero (as $j \to \infty$), it follows from the foregoing that an arbitrary surface of type (g, n), which (as we have seen) is a subregion of R_k, is also a subregion of the infinite set of surfaces $R_{k,j}$ (only one would suffice for us).

It remains to put the surfaces $R_{k,j}$ on a single universal surface. To do this, take a plane and on it countably many disjoint disks, and glue the surfaces $R_{k,j}$ along the circles bounding these disks. For complete universality it is necessary to repeat this procedure for each genus $g = 1, 2, \ldots$.

EXAMPLE 78 (GREENBERG [89]). *A regular covering $\tilde{S} \to S$ over a given Riemann surface S with a given finite group $G(\tilde{S}, S)$ of covering transformations that coincides with* $\operatorname{Aut} \tilde{S}$. Let S be a compact Riemann surface of genus g, and let G be a finite nontrivial group. Choose a set of generators $x_1, \ldots, x_k$ for G such that: a) x_i has order $\nu_i > 1$, $i = 1, \ldots, k$; b) $x_1 x_2 \cdots x_k = 1$; and c) $k > 6g - 3$ and $k > 6$ if $g \leq 1$. This choice is always possible by adding elements of the form x, x^{-1} to the list of generators, where $x \in G \backslash \{1\}$.

Greenberg [89] established that a Riemann surface S can be uniformized by a maximal Fuchsian group Γ (one not contained in another Fuchsian group) of signature $(g, k; v_1, \ldots, v_k)$ with sufficiently many elements of finite order, i.e., in the case when the condition c) holds. Let $S = U/\Gamma$, and suppose that Γ is generated by elements a_i, b_i, c_j $(1 \leq i \leq g, 1 \leq j \leq k)$ with defining relations

$$\prod_{i=1}^{g} [a_i, b_i] \prod_{j=1}^{k} c_j = 1, \quad c_j^{\nu_j} = 1, \qquad 1 \leq j \leq k. \tag{91}$$

We define an epimorphism of Γ onto G by the rule

$$\begin{aligned} \theta(a_i) = \theta(b_i) = 1, \qquad & i = 1, \dots, g, \\ \theta(c_j) = x_j, \qquad & j = 1, \dots, k. \end{aligned} \tag{92}$$

Let $\Gamma_0 = \operatorname{Ker}\theta$. Since c_j and x_j have the same order, (91) and (92) give us that Γ_0 does not contain elements of finite order, and $\Gamma/\Gamma_0 \cong G$. Let $\tilde{S} = U/\Gamma_0$, and let $p\colon \tilde{S} \to S$ be the regular covering induced by the group inclusion $\Gamma_0 \subset \Gamma$. For the group of covering transformations of the covering p we have the isomorphisms

$$G(\tilde{S}, S) \cong N(\Gamma_0, \Gamma)/\Gamma_0 \cong \Gamma/\Gamma_0 \cong G. \tag{93}$$

Let us now compute the full group of conformal automorphisms of $\tilde{S}$. To do this, observe that $\Gamma \subset N(\Gamma_0, \operatorname{Aut} U)$ and is a maximal Fuchsian group. Consequently,

$$\operatorname{Aut}\tilde{S} \cong N(\Gamma_0, \operatorname{Aut} U)/\Gamma_0 \cong \Gamma/\Gamma_0 \cong G. \tag{94}$$

Comparing (93) and (94), we get the required property of the covering p, namely, that $G(\tilde{S}, S) \cong \operatorname{Aut}\tilde{S} \cong G$.

EXAMPLE 79 (ACCOLA [10]). *A compact Riemann surface with trivial group of conformal automorphisms.* Let S be a compact Riemann surface of genus g on which a finite set E is singled out such that the identity is the only conformal automorphism of S leaving E invariant. Let $\Phi\colon \tilde{S} \to S$ be a p-sheeted branched covering with branching set E and with trivial covering transformation group, such that $r_b > 2p(p-1)$, where r_b is the sum of all the ramification indices of the covering Φ. If p is prime and greater than 2, then Φ is a maximal strictly branched covering in the sense of Accola [10] and has the following properties:

a) $G_\Phi(\tilde{S}, S)$ is a central subgroup of $\operatorname{Aut}\tilde{S}$.

b) Each conformal automorphism of $\tilde{S}$ can be pushed down to a conformal automorphism of the surface S.

c) $\operatorname{Aut}\tilde{S}/G_\Phi(\tilde{S}, S) \subset \operatorname{Aut}_E S$, where $\operatorname{Aut}_E S$ is the set of conformal automorphisms of S leaving E invariant.

By the choice of the set E, the groups $\operatorname{Aut}_E S$ and $G_\Phi(\tilde{S}, S)$ are trivial, and this implies that the group $\operatorname{Aut}\tilde{S} = \operatorname{Aut}\tilde{S}/G_\Phi(\tilde{S}, S)$ is trivial.

The genus $\tilde{g}$ of $\tilde{S}$ is determined from the Riemann-Hurwitz relation $2\tilde{g}-2 = p(2g-2) + r_b$, and is minimal when $g = 0, p = 3$, and $r_b = 14$. In this case $\tilde{g} = 5$. To construct a surface $\tilde{S}$ of genus $\tilde{g} = 5$ we can take as E any set of eight points in the extended complex plane $S = \overline{\mathbf{C}}$ that is not left invariant under any linear fractional mapping except the identity. We require that the three-sheeted covering Φ have branching order 2 over two of the points in E and branching order 3 over the others. Moreover, Φ is not a covering of regular type, and hence $G_\Phi(\tilde{S}, S) = \{1\}$. The condition $r_b > 2p(p-1)$ can be verified immediately.

This example does not give an analytic description of the Riemann surface. Such a description in terms of the corresponding Fuchsian group is given in the next example, where a whole class of surfaces with an analogous property is constructed.

EXAMPLE 80 (MEDNYKH [191]). *Fuchsian groups that uniformize compact Riemann surfaces with trivial group of conformal automorphisms.* Let $U = \{z \in \mathbf{C}\colon |z| < 1\}$ be the unit disk, which we take as a model of the Lobachevsky plane with the metric $d\sigma = |dz|/(1-|z|^2)$. For a given prime number $p > 3$ we define three permutations on p-symbols:

$$\begin{aligned} \varsigma &= (p-1, p-2, \dots, 4, 3, 1, p, 2), \\ \eta &= (1, 4, 6, \dots, p-1, 3, 5, p, 2), \\ \xi &= (1, 2, \dots, p-1, p), \end{aligned} \tag{95}$$

which satisfy the condition $\varsigma\eta = \xi$. Further, suppose that we are given r cycles of length p such that

$$\theta_1\theta_2\cdots\theta_r = \xi, \tag{96}$$

where r is an integer such that $r \geq 2p$.

In the disk U we construct a non-Euclidean triangle OAB whose vertex O is at the point $z = 0$, while the angles at the vertices O, A, and B are equal to $\pi/pr, \pi/p$, and π/r, respectively. This triangle is denoted by F, and its side AB by I. The triangle and side corresponding to them under reflection with respect to the side OA are denoted by F^- and I^-. Let $y = e^{2\pi i/pr}z$ be an elliptic transformation of U having order pr. Consider the non-Euclidean polygon

$$F_0 = \bigcup_{l=0}^{pr-1} y^l(F \cup F^-), \tag{97}$$

which has $2pr$ sides (see Figure 60) determined in terms of the sides of F and F^- by the formulas

$$\begin{aligned} I^+_{k,j} = y^{jr+k-1}(I), \quad I^-_{k,j} &= y^{jr+k-1}(I^-), \\ k = 1, 2, \dots, r, \ j &= 1, 2, \dots, p. \end{aligned} \tag{98}$$

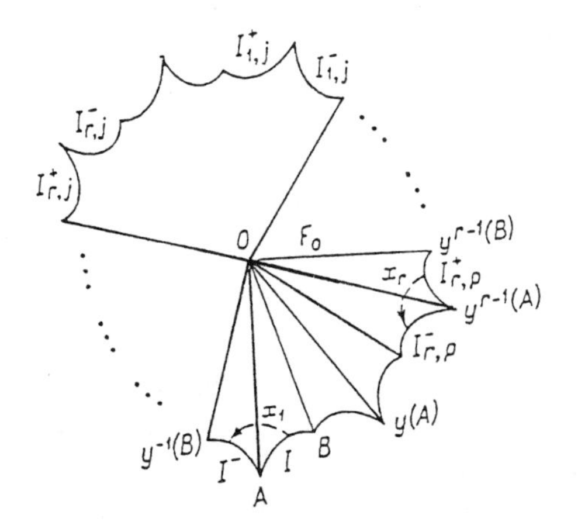

FIGURE 60

Denote by x_k the elliptic transformations of order p mapping the side $I^+_{k,p}$ onto $I^-_{k,p}$, $k = 1, 2, \dots, p$. Mappings that identify the sides (98) of F_0 pairwise are given with the help of permutations in (95) and (96):

$$T_{k,j} = y^{r\theta_k(j)} x_k y^{-rj}, \qquad k = 1, \dots, r;\ j = 1, \dots, p. \tag{99}$$

Note that $T_{k,j}$ maps the side $I^+_{k,j}$ onto $I^-_{k,\theta_k(j)}$. By the Poincaré theorem (see Problem 9), the transformations (99) generate a Fuchsian group Γ_0 for which (97) is a fundamental polygon. Moreover, Γ_0 does not contain elliptic elements and determines a compact Riemann surface $T = U/\Gamma_0$.

We now indicate an explicit form for the permutations in (96) corresponding to a Riemann surface $T = U/\Gamma_0$ with trivial group of conformal automorphisms. For even r let

$$\begin{aligned} \theta_k &= \xi^{(-1)^k}, \qquad k = 1, \dots, r-2; \\ \theta_{r-1} &= \varsigma, \qquad \theta_r = \eta. \end{aligned} \tag{100}$$

For odd r let

$$\begin{aligned} \theta_k = \xi^{(-1)^k}, \quad k = 1, \dots, r-4; \qquad \theta_{r-3} = \xi^2, \\ \theta_{r-2} = \xi^{-1}, \qquad \theta_{r-1} = \varsigma, \qquad \theta_r = \eta. \end{aligned} \tag{101}$$

It can be proved (see [191, Theorem 1]) that if p is a prime number > 3, $r \geq 2p$, and T is a Riemann surface determined by the conditions (100) or (101), then $\operatorname{Aut} T \cong \{1\}$.

EXAMPLE 81 (MEDNYKH [191]). *A compact Riemann surface S with $\operatorname{Aut} S$ a cyclic group.* Let $S = \overline{\mathbf{C}}$ and let N be an integer ≥ 2. Further, let p be a prime number such that $p \equiv 1 \pmod{2N^2}$. We show that there exist a compact Riemann surface $\tilde{S}$ and a p-sheeted covering $\Phi\colon \tilde{S} \to S$ of regular type such that $G(\tilde{S}, S) \cong \{1\}$ and $\operatorname{Aut} \tilde{S} = \mathbf{Z}_N$, where $\mathbf{Z}_N$ is the cyclic group of order N. To do this, take $p = 2m_0N^2 + 1$ ($m_0 \geq 1$), the permutation $\xi = (1, 2, \ldots, p)$, and a cycle τ of length p such that $(\xi^{2m_0N}\tau)^N = 1$. Denote by i_k the remainder after division of k by p. In the notation of the preceding example we set $r = N_p$ and consider the permutations

$$\begin{aligned} &\theta_k = \xi^{(-1)^{i_k}}, \qquad p \text{ does not divide } k,\ 1 < k < r, \\ &\theta_{l_p}\xi^{(l-1)m}\tau\xi^{-(l-1)m}, \qquad l = 1, 2, \ldots, N. \end{aligned} \tag{102}$$

The condition (96) on the permutation (102) can be verified directly. From the construction of the preceding example we get that $\tilde{S} = U/\Gamma_0$ and $S = U/\Gamma$, where $\Gamma = \langle y, \Gamma_0 \rangle$ is the group generated by y and Γ_0. It follows from Lemmas 1 and 2 in [191] that $N(\Gamma_0, \Gamma) = \Gamma_0$ and

$$N(\Gamma_0, \operatorname{Aut} U) = \sum_{l=0}^{N-1} y^{lp^2}\Gamma_0. \tag{103}$$

The properties of $\operatorname{Aut} \tilde{S}$ and formula (103) give us that $\operatorname{Aut} \tilde{S}$ (which is isomorphic to the quotient of $N(\Gamma_0, \operatorname{Aut} U)$ modulo Γ_0) is isomorphic to the cyclic group $\mathbf{Z}_N$, while the covering transformation group $G(\tilde{S}, S) \cong N(\Gamma_0, \Gamma)/\Gamma_0$ is trivial.

EXAMPLE 82. *The n-dimensional torus.* The n-dimensional torus T^n is defined as the Cartesian product of n circles. This n-dimensional manifold can be uniformized by an elementary Kleinian group G—the free abelian group of rank n with generators the translations in $\mathbf{R}^n$ by the vectors $e_k = (0, \ldots, 0, 1, 0, \ldots, 0)$, $k = 1, \ldots, n$; a cube with edges of unit length is a fundamental region for G. The manifold $T_n = \mathbf{R}^n/G$ is obtained by identifying opposite faces of the cube.

The following examples have to do with the construction of conformally Euclidean manifolds with infinite nonbelian fundamental groups, and they show that the structure of such manifolds can be very complicated in the general case.

EXAMPLE 83 (APANASOV). *The n-dimensional sphere with p handles.* As is known, any closed Riemann surface of genus p can be represented as a connected sum of the two-dimensional sphere and p copies of two-dimensional tori. Similarly, we can consider a closed compact n-dimensional manifold M_p^n that is a connected sum of the n-dimensional sphere and p copies of the n-dimensional tori T^n described in Example 82. For $n \geq 3$ the fundamental group $\pi_1(M_p^n)$ (see Figure 61) has the presentation

$$\{a_1, b_1, c_1, \ldots, a_p, b_p, c_p \colon [a_1, b_1] = [b_1, c_1] = \cdots = [c_p, a_p] = 1\}. \tag{104}$$

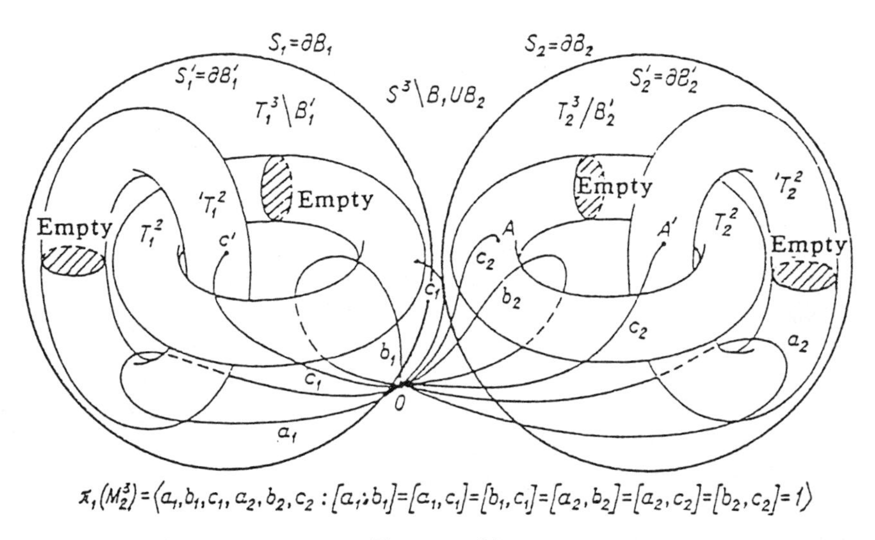

FIGURE 61

The manifold M_p^n can be uniformized by a Kleinian group $G \subset \mathcal{M}_n$. As such a group it suffices to take the free product of p free abelian groups of rank n, i.e., a Kleinian group G having the presentation (104) and obtained by Klein combination of p elementary groups $G_1, \dots, G_p$, each conjugate in $\mathcal{M}_n$ to a group generated by translations by vectors in general position.

Thus, the n-sphere with p handles $M_p^n = \Omega(G)/G$ is covered by the region $\Omega(G) \subset \mathbf{R}^n$ obtained from $\mathbf{R}^n$ by deleting the perfect nowhere dense set $\Lambda(G)$, which has positive logarithmic capacity. In a way similar to that in Example 31 it can be shown that, for a sufficiently large number of handles, the set $\Lambda(G)$ has Hausdorff dimension greater than $n/2$.

The region $\Omega(G)$ is the universal covering space of the conformally Euclidean manifold M_p^n; consequently, the n-sphere with p handles is not a hyperbolic manifold when $n \geq 3$ (contrary to the case $n = 2$).

EXAMPLE 84 (KUIPER [144]). Consider in $\mathbf{R}^n$ the two-dimensional half-plane $H = \{(x_1, \dots, x_n) \in \mathbf{R}^n \colon x_1 > 0, x_2 = \dots = x_{n-1} = 0\}$ with the hyperbolic metric $ds^2 = (dx_1^2 + dx_2^2)/x_1^2$. The geodesics in this metric are represented by arcs of circles orthogonal to the line $x_1 = 0$. Let P be a convex hyperbolic hexagon in H with right angles. If we perform successively all the inversions J with respect to the sides of P and the resulting images of them, then H is entirely filled by the images of P. Identification of equivalent (with respect to the inversions) points of H gives a closed two-dimensional Riemann manifold of constant negative curvature.

Now rotate H about the line $l \colon x_1 = \dots = x_{n-1} = 0$. This gives the whole of $\mathbf{R}^n$ less the line l. The inversions J considered above uniquely determine corresponding inversions $\tilde{J}$ in $\mathbf{R}^n$. These inversions commute with the rotations of $\mathbf{R}^n$ about l. Identifying points of $\mathbf{R}^n \setminus l$ that are equivalent with respect to an even number of inversions J, we get a compact conformally Euclidean manifold; in this manifold we can define a global metric

$$ds^2 = \sum_{j=1}^{n} dx_j^2 \Big/ \sum_{j=1}^{n-1} x_j^2$$

(which is not hyperbolic).

The following constructions permit us to essentially broaden the class of conformally Euclidean manifolds with infinite nonabelian fundamental groups.

EXAMPLE 85. Let Γ be an arbitrary finitely generated torsion-free Fuchsian group of the first kind in $\overline{\mathbf{C}}$ with limit circle l. We extend the action of Γ to the whole space $\overline{\mathbf{R}}^n$; the limit set of the extended group also coincides with l. The manifold $M = (\overline{\mathbf{R}}^n \backslash l)/\Gamma$ is conformally Euclidean and is a trivial fibering over the circle with fiber the Riemann surface $S = (\operatorname{int} l)/\Gamma$, which is uniformizable by a group in the disk. The group $\pi_1(M)$ is automatically infinite, since it contains an infinite cyclic subgroup isomorphic to $\pi_1(\mathbf{R}^3 \backslash l)$, and it is nonabelian (in view of what was said in §8 of Chapter II).

Moreover, if Γ does not contain parabolic elements, then M is a compact manifold.

EXAMPLE 86. Let G be a torsion-free Kleinian group in $\overline{\mathbf{C}}$ that is a $\mathbf{Z}_2$-extension of a finitely generated quasi-Fuchsian group without parabolic elements (see Examples 19 and 27). Then (as shown by Gusevskiĭ [92]) G is isomorphic to the fundamental group of a closed nonorientable surface S, and the manifold $M^+(G) = (\mathbf{R}^3_+ \cup \Omega(G))/G$ is a nontrivial linear fibering over S. The manifold $M^-(G) = (\mathbf{R}^3_- \cup \Omega(G))/G$ obviously has the same structure. These manifolds have the common boundary $\Omega(G)/G$, which represents a single (orientable) surface. Therefore, if it is assumed that the action of G is extended to the whole of $\overline{\mathbf{R}}^3$, then the conformally Euclidean manifold $M(G) = (\overline{\mathbf{R}}^3 \backslash L)/G$, where L is the limit Jordan curve of G in $\overline{\mathbf{C}}$, is obtained by gluing $M^+(G)$ and $M^-(G)$ along their common boundary, and is a nontrivial fibering into circles over the original nonorientable surface S.

EXAMPLE 87 (APANASOV [37]). *A 3-manifold covered by the space of a wild knot.* Fix a finite family of spheres S_i $(i = 1, \dots, k)$ and some polygonal knot l in $\mathbf{R}^3$ whose vertices are points of tangency of the S_i (each sphere is tangent from the outside to exactly two others). Consider the $(k-1)$-generated Kleinian group $G \subset \mathcal{M}_3$ consisting of superpositions of an even number of inversions with respect to the spheres S_i. Then the spherical polyhedron

$$P(G) = \bigcap \operatorname{ext} S_i \cup J_k\left(\bigcap \operatorname{ext} S_i\right) \tag{105}$$

is a fundamental polyhedron for G. The points of tangency of its sides (which are parabolic vertices) are the vertices of a polygonal knot l_0 having the same type as the knot composition $l \# l$ (see [69], and also (106)). It is easy to see that the limit set $\Lambda(G)$ forms a simple closed curve passing through the vertices of $P(G)$ and their G-images. The type of this knot becomes clear if we observe that the inversion J with respect to the sphere S bounding the polyhedron (105) yields a knot l_1 having the type of the knot composition $l_0 \# l_0$, which can be represented as the union

$$(l_0 \cap \operatorname{ext} S) \cup (J(l_0) \cap \operatorname{int} S). \tag{106}$$

Continuing this process ad infinitum, we obtain a sequence $\{l_m\}$ of knots such that l_m has the type of the knot $l_{m-1} \# l_{m-1}$ and lies in the complement of the union of $[m/2]+1$ images of the polyhedron (105). As the set $\Omega(G)$ is exhausted by the images of $P(G)$, this sequence of knots approximates the limit set $\Lambda(G)$, which is thus a wild knot in $\mathbf{R}^3$ that does not have a single handle-arc (see [69]).

Accordingly, we have constructed a Kleinian group G that uniformizes the 3-manifold $\Omega(G)/G$, whose covering space $\Omega(G)$ is the space of a wild knot $\Lambda(G)$. The fact that the knot thus obtained is wild arises from necessity. It is known [145] that the limit set $\Lambda(G)$ of a Kleinian group $G \subset \mathcal{M}_3$ cannot be a tame knot.

The manifold $\Omega(G)/G$ obtained is quasiconformally equivalent to a manifold of finite type (the space of orbits of a finitely generated Fuchsian group similar to the group in Example 85). Indeed, consider in the plane $\mathbf{C}$ a covering of the circle S^1 by disks b_i $(i=1,\dots,k)$ (as in the case of the spheres S_i, the disks are tangent to each other, and their boundaries are orthogonal to S^1); in $\mathbf{R}^3$ take a family of spheres $\mathcal{S}_i$ $(i=1,\dots,k)$ with the same centers and radii as the disks b_i, and form the $(k-1)$-generated Kleinian group $\Gamma \subset \mathcal{M}_3$ consisting of superpositions of an even number of inversions with respect to the spheres $\mathcal{S}_i$. The group Γ is the extension of a Fuchsian group of finite type from $\mathbf{C}$ to $\mathbf{R}^3$.

Considering now the cones

$$\{x \in \mathbf{R}^4 \colon x = \lambda y + \mu p,\ y \in l_0\}, \quad \{x \in \mathbf{R}^4 \colon x = \lambda y + \mu q,\ y \in S^1\}, \tag{107}$$

where the numbers λ and μ are positive, $\lambda+\mu=1$, and p and q are particular points in the half-space $\mathbf{R}^4_+$, we can use methods from piecewise linear topology (extension of a mapping carrying one of the cones in (107) into the other; see [212]) to construct a quasiconformal homeomorphism f_0 of $\mathbf{R}^4$ carrying the knot l_0 into S^1, and the fundamental polyhedron $P(\hat{G})$ into the fundamental polyhedron $P(\hat{\Gamma})$ with correspondence of sides. Here $\hat{G}$ and $\hat{\Gamma}$ denote the extensions to $\overline{\mathbf{R}}^4$ of the Kleinian groups G and Γ, and $P(\hat{G})$ and $P(\hat{\Gamma})$ denote the polyhedra analogous to (105). If we now extend f_0 (in a way compatible with the actions of $\hat{G}$ and $\hat{\Gamma}$) from $P(\hat{G})$ to the set $\Omega(\hat{G})$ of discontinuity, and then to the removable set $\Lambda(G)=\Lambda(\hat{G})$ (see [37] for details), then we get a quasi-conformal automorphism f of $\overline{\mathbf{R}}^4$ such that $f\hat{G}g^{-1}=\hat{\Gamma}$. This automorphism f "unties" the wild knot $\Lambda(G), f(\Lambda(G)) = f(\Lambda(\hat{\Gamma})) = f(\Lambda(\Gamma)) = S^1$, and the inverse homeomorphism f^{-1} induces a quasiconformal mapping of the

4-manifold $\Omega(\hat{\Gamma})/\hat{\Gamma}$ of finite type onto the manifold $\Omega(\hat{G})/\hat{G}$, which is decomposed by the 3-manifold $\Omega(G)/G$ into two identical manifolds with boundary.

EXAMPLE 88 (THURSTON [234]). *Manifolds having nonhyperbolic structure and obtainable by Dehn surgery from a noncompact hyperbolic manifold.* As a manifold M with a complete hyperbolic structure let us consider the complement in S^3 of the Borromean rings (see Examples 60, 67, and 37). Trivial Dehn surgery with the parameter $d_1 = (1,0)$ with respect to one of the link circles turns M into a manifold $M_{(1,0),\infty,\infty}$ homeomorphic to S^3 with two nonlinked circles removed. After this, any Dehn surgery of the resulting manifold $M_{(1,0),\infty,\infty}$ with parameters d_2 and d_3 (d_2 and d_3 are primitive elements of $H_1(T_i^2, Z)$, $i = 2, 3$, and the T_i^2 are two-dimension tori) gives only a connected sum of lens spaces (see Example 72 and Exercise 134), i.e., a manifold homeomorphic to $\Omega(G)/G$, where $G \subset \mathcal{M}_3$ is the nonelementary Kleinian group obtained in the separate cases by free combination of the finite groups G_1 and G_2. It is easy to see that of the manifold $\Omega(G)/G = M_{(1,0),d_2,d_3}$ is never hyperbolic. The region $\Omega(G) \subset S^3$ is its covering space.

Thus, a frequently encountered case is that in which Dehn surgery is used to make a complete hyperbolic manifold into an infinite set of manifolds having nonhyperbolic structure.

EXAMPLE 89 (MASKIT [188]). *Isomorphic fundamental groups with isomorphic factor subgroups that are not topologically equivalent.* The concept of the signature of a finitely generated fundamental group G is presented in detail in Problem 41. Here we mention only that the signature of G is a pair (p, K), where p is the genus of the surface $S_0 = \Delta/G$ (Δ is an invariant component of G), and K is a two-dimensional complex.

Let S be a Riemann sphere with five marked points $x_1, \dots, x_5$. Let v_1 be a simple loop on S that separates x_1 and x_2 from x_3, x_4, x_5; v_2 is a simple loop disjoint from v_1 and separating x_1, x_2, x_3 from x_4 and x_5. Using results in §4 of Chapter II, we construct a Kleinian group G_1 as the group of covering homeomorphisms of a branched covering $\pi_1 \colon \Delta_1 \to S$ for which v_1 and v_2 lift to loops in an invariant component Δ_1, while the points $x_1, \dots, x_5$ are branch points of respective orders $2, 3, 5, 7, 2$. Then G_1 has signature $(0, K_1)$, where the complex K_1 is pictured in Figure 62.

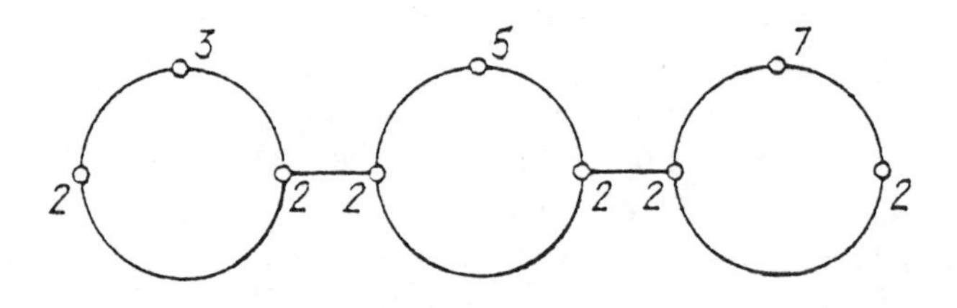

FIGURE 62

The group G_1 has three classes of conjugate factor subgroups with respective uniformization signatures $(0,3;2,2,3),(0,3;2,2,5)$ and $(0,3;2,2,7)$ (see §4 of Chapter II). Maskit combination can be used to obtain G_1 from elementary groups with these signatures (G_1 can be constructed successively as the free product with amalgamated subgroup of order 2). It has the following presentation:

$$G_1 = \{A, B, C, D\colon A^2 = B^2 = (AB)^2 \\ = C^2 = (BC)^5 = D^2 = (CD)^7 = 1\}.$$

The three nonconjugate factor subgroups of G_1 are the groups generated by A and B, B and C, and C and D.

We construct a group G_2 similarly, with the single difference that for it we interchange the branching numbers 5 and 7. Then G_2 has the signature $(0, K_2)$, where K_2 is the complex pictured in Figure 63.

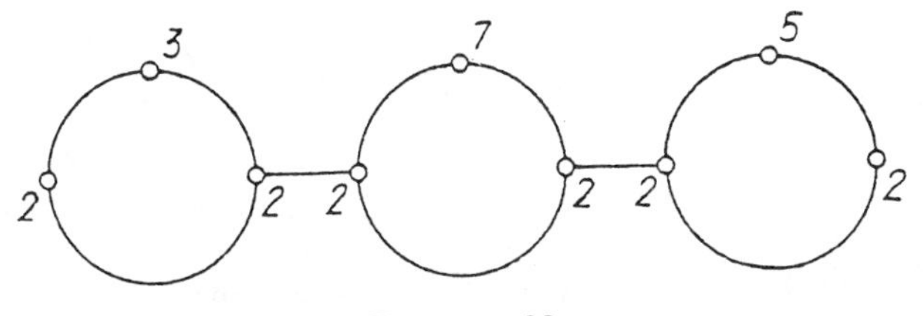

FIGURE 63

Since G_1 and G_2 have different signatures, these groups are not similar (see Problem 41). Moreover, if their actions are extended into $\mathbf{R}^3_+$, then there is no homeomorphism of $\mathbf{R}^3_+$ that conjugates these groups.

Considering that G_2 has the presentation

$$G_2 = \{a, b, c, d\colon a^2 = (ab)^3 = c^2 = (bc)^7 = d^2 = (cd)^5 = 1\}, \tag{109}$$

we define a homomorphism $\psi\colon G_1 \to G_2$ by

$$\begin{aligned} &\psi(A) = (cd)^2(bc)^3a(cb)^3(dc)^3, \\ &\psi(B) = d, \qquad \psi(C) = c, \qquad \psi(D) = b. \end{aligned} \tag{110}$$

It follows from (108)–(110) that ψ is an isomorphism. Considering that each elliptic element of G_1 lies in some factor subgroup of G_1, we get that ψ and ψ^{-1} preserve type, i.e., they are admissible in the sense of §5 in Chapter I. Therefore, the factor subgroups of G_1 and G_2 are isomorphic.

EXAMPLE 90 (GUSEVSKIĬ). *A "nice" three-dimensional manifold that cannot be uniformized by any Kleinian group.* Let $M_0 = S \times [0,1]$, where S is a torus. We identify the two topological disks $D_1 \subset S \times \{0\}$ and $D_2 \subset S \times \{1\}$ by means of an orientation-preserving homeomorphism. The manifold thus obtained (see Figure 64) is denoted by M. It is not hard to verify that M is a compact orientable aspherical irreducible manifold whose boundary is a surface of genus 2. By van Kampen's theorem, $\pi_1(M)$ is the free product of a free abelian group of rank 2 and an infinite cyclic group, and it thus has trivial center.

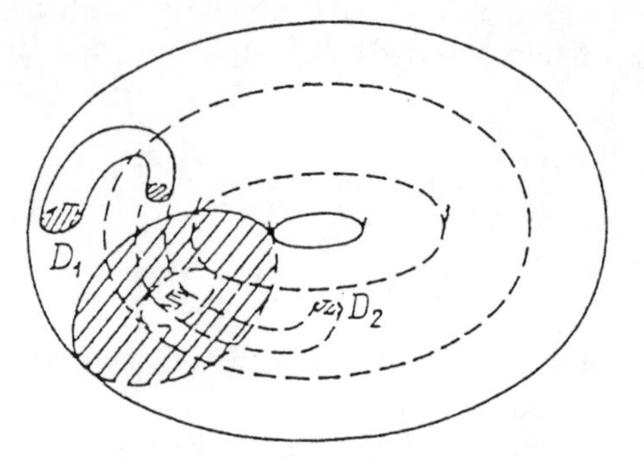

FIGURE 64

We show that M cannot be uniformized by a Kleinian group, i.e, M is not homeomorphic to the manifold $M(G)$ for any Kleinian group G. Indeed, otherwise G would contain a free abelian subgroup of rank 2, which is known to be generated always by parabolic mappings in the case of Kleinian groups. But then $M(G)$ contains a submanifold homeomorphic to a punctured torus $(U\backslash\{0\}) \times S^1$, and this cannot be, because M is compact.

Now let N be a disk sum of $S\times[0,1)$ and $\overline{U}\times S^1$. Then the manifold N can be uniformized by a group of Schottky type obtained by Klein combination of a free abelian group of rank 2 (a group of translations in two directions) and a cyclic loxodromic group.

We have thereby constructed two orientable aspherical irreducible manifolds with connected boundary whose fundamental groups are isomorphic and have trivial centers, and one of them can be uniformized by a Kleinian group while the other cannot.

CHAPTER IV

Problems and Exercises
(further development of the theory)

§1. Exercises on the geometry of Kleinian groups

1. Show that two linear fractional transformations γ_1 and γ_2, different from the identity, commute if they have the same fixed points. Conversely, if two transformations $\gamma_1, \gamma_2 \neq I$ commute, then they either have the same fixed points or are elliptic transformations of order two.

2. Find necessary and sufficient conditions for two Möbius mappings in $\mathbf{R}^n$ to commute when $n > 2$.

3. Prove that a discrete group $G \subset \mathcal{M}_n$ is at most countable.

4. Show that a Kleinian group $G \subset \mathcal{M}_n$ is discrete (the converse is not true; see Example 15).

5. Show that the extension of any discrete group $G \subset \mathcal{M}_n$ to the half-space $\mathbf{R}^{n+1}_+$ is a Kleinian group.

6. Show that the fixed points of the hyperbolic and loxodromic elements and the parabolic elements of a nonelementary Kleinian group $G \subset \mathcal{M}_n$ are dense in their limit set.

7. Show that a discrete group $G \subset \mathcal{M}_n$ consisting of elliptic elements (besides the identity) is finite.

8. Describe all the finite groups of Möbius mappings.

9. Prove that if a Kleinian group $G \subset \mathcal{M}_n$ contains only parabolic transformations, then these transformations have a common fixed point; for $n = 2$ this group is either discrete or free abelian of rank 2.

10. Describe all the abelian discontinuous groups of Möbius mappings.

11. Prove that if a Kleinian group $G \subset \mathcal{M}_n$ consists only of parabolic and elliptic elements and they all have a common fixed point, then this point is the only limit point of the group.

12. Prove that a Kleinian group with one limit point is a finite extension of a cyclic parabolic group or of a free abelian group of rank 2, while a Kleinian group with two limit points is a finite extension of a cyclic loxodromic group (cf. Exercise 136).

13. Prove that if the limit set of a Kleinian group $G \subset \mathcal{M}_n$ consists of more than one point, then G contains hyperbolic or loxodromic elements.

14. Show that if G is a group of Möbius transformations, $\gamma_1 \in G$ is of loxodromic type, and $\gamma_2 \in G$ has a single common fixed point with γ_1, then the group is not discrete.

In particular, a parabolic element of a discontinuous group cannot have a common fixed point with a loxodromic (hyperbolic) element of it. The preceding exercises give that a discontinuous group whose elements have a common fixed point can have at most two limit points, i.e., is elementary.

15. Show that a Kleinian group $G \subset \mathcal{M}_2$ with more than two limit points is Fuchsian if and only if it does not contain loxodromic elements.

16. Show that a Kleinian group $G \subset \mathcal{M}_2$ can have one or two or countably many components; moreover, each component of it is simply connected or doubly connected or infinitely connected (as Example 22 shows, that is not true for spatial groups).

17. Show that a Fuchsian group in the disk with a compact fundamental region consists solely of hyperbolic and, possibly, elliptic elements. The corresponding assertion is also valid for the multi-dimensional case.

18. Show that if G_0 is a subgroup of a Kleinian group $G \subset \mathcal{M}_n$ of finite index, then $\Lambda(G_0) = \Lambda(G)$; the converse is false.

19. Let G_0 be a subgroup of a geometrically finite Fuchsian group $G \subset \mathcal{M}_n$ such that $\Lambda(G_0) = \Lambda(G)$. Must G_0 have finite index in G?

20. Show that if G_0 is a nonelementary normal subgroup of a Kleinian group $G \subset \mathcal{M}_n$, then $\Lambda(G_0) = \Lambda(G)$.

21. Show that if a closed set $S \subset \overline{\mathbf{R}}^n$ containing more than two points is invariant under a discrete group $G \subset \mathcal{M}_n$, then S contains the limit set $\Lambda(G)$.

22. Show that the normalizer of a group $G \subset \mathcal{M}(2)$ in $\mathcal{M}(2)$ is discrete if and only if G itself is discrete.

This is not true in dimensions $n > 2$; give a corresponding example.

23. Show that if $G \subset \mathcal{M}_n$ is a nonelementary Kleinian group, then:

(a) each limit point of G is an accumulation point of the images of any point in $\overline{\mathbf{R}}^n$; and

(b) if $x_0 \in \Omega(G)$ and $\{g_k\}$ is a sequence of distinct elements of G such that $g_k(x_0) \to x^1$, then there exists a subsequence $\{g_{k_j}\}$ such that $\lim_{j\to\infty} g_{k_j}(x) = x^1$ for all but possibly one point $x \in \overline{\mathbf{R}}^n$.

24. Let Γ be a Fuchsian group acting in the disk U. Prove that the following conditions are equivalent:

(a) There exists a fundamental region F of Γ in U of finite non-Euclidean area.

(b) The non-Euclidean area of any (measurable) fundamental region F of Γ in U is finite.

(c) A normal fundamental polygon P of Γ in U has finitely many sides and does not have free sides (i.e., sides that are nondegenerate arcs of the circle ∂U).

(d) Γ is a finitely generated group of the first kind.

(e) The Riemann surface U/Γ has finite type.

Conditions (a)–(c) remain in force also for Fuchsian groups acting in the ball $B^n, n > 2$.

25. Using the properties of a hyperbolic metric, prove the following assertion:

AHLFORS' LEMMA. *Let Δ be a component of a Kleinian group G, and let Δ/G_Δ be a Riemann surface S with deleted point p_0 such that the projection $\pi\colon \Delta \to \Delta/G_\Delta$ is not branched over S in some deleted neighborhood of p_0 on S. Then there exist a parabolic element $A \in G$ with fixed point $z_0 \in \Lambda(G)$ and a linear fractional transformation B such that* 1) $B(\infty) = z_0$ *and $B^{-1} \circ A \circ B$ is the translation $z \to z+1$;* 2) $B^{-1}(\Delta)$ *contains a half-plane $H_c = \{z \in \mathbf{C}\colon \operatorname{Im} z > c\}$, and two points $z_1, z_2 \in B(H_c)$ are equivalent with respect to G if and only if $z_2 = A^n(z_1)$ for some integer n; and* 3) $\pi[B(H_c)]$ *is a deleted neighborhood of p_0 on S that is homeomorphic to the punctured disk* $0 < |\varsigma| < 1$.

The region $B(\{z \in \mathbf{C}\colon 0 < \operatorname{Re} z < 1, \operatorname{Im} z > c\})$ is called the *parabolic region* (the triangle) *belonging to* p_0, and $B(H_c)$ is called the *half-plane belonging to* p_0.

26. Let $G \subset \mathcal{M}_n$ be a Kleinian group, $G_\infty \subset G$ the subgroup of G preserving ∞, and F a fundamental region of G_∞ bounded by pairwise equivalent sides. Prove that its intersection with the exterior of all the isometric spheres of elements of G forms a fundamental region of G.[15]

27. Show that a discrete group G of isometries of an n-dimensional simply connected complete space X of constant curvature is generated by the set $O = \{g \in G\colon \dim(P \cap g(P)) = n-1\}$, where $P \subset X$ is a convex fundamental polyhedron of G. Moreover, the defining relations of G can be taken to be all possible edge relations of the following two types: $g_1g_2 = \mathbf{1}$, where $g_1, g_2 \in O$; $g_1g_2, \dots, g_k = 1$, where $g_1, \dots, g_k \in O$; $\dim\{P \cap g_1(P) \cap g_1g_2(P) \cap \dots \cap g_1g_2\cdots g_{k-1}(P)\} = n-2$, $g_ig_{i+1} \neq \mathbf{1}$ for $i = 1, \dots, k-1$, and $g_1g_2\cdots g_m \neq \mathbf{1}$ for $m < k$.

28. A group G of Möbius automorphisms of $\overline{\mathbf{R}}^n$ $(n \geq 2)$ is said to be a *generalized Schottky group* if its limit set is totally disconnected (i.e., does not contain a continuum different from a point).

Show that if G is a finitely generated torsion-free generalized Schottky group, then:

1) G is isomorphic to a free product $G_0 * G_1 * \cdots * G_k$, $k < \infty$, where G_0 is a free group and G_j $(j = 1, \dots, k)$ is conjugate to a group of Euclidean

---

[15] *Translation editor's note*: This might not be true if G_∞ is finite.

motions; and

2) G can be obtained from elementary groups by Klein combination.

29. Following Maskit [181], we define a *conglomerate* to be a collection (G, D, H, Δ, S) of objects with G a Kleinian subgroup of $\mathcal{M}_n$ with fundamental set D, H a subgroup of G with fundamental set Δ, and S the Jordan surface bounding an n-dimensional ball B such that a) $D \subset \Delta$; b) S is invariant under H, and c) there exists a neighborhood V of S for which $\Delta \cap V \subset D$.

A *large conglomerate* (G, D, H, Δ, S, B) is defined to be a conglomerate such that B $(\partial B = S)$ is invariant under H and d) $\Delta \cap (B \cup S) = D \cap (B \cup S)$, e) $D \cap (B \cup S) \neq D$, and f) $g(S) \cap S = \varnothing$ for all $g \in G \backslash H$.

Using Maskit's arguments [181], prove the following combination theorems:

(I) *If* $(G_1, D_1, H_1, \Delta, S, B_1)$ *and* $(G_2, D_2, H, \Delta, S, B_2)$ *are large conglomerates,* $B_1 \cap B_2 = \varnothing$, *and the group* $G \subset \mathcal{M}_n$ *is generated by* G_1 *and* G_2, *then* G *is Kleinian with fundamental set* $D = D_1 \cap D_2$ *and is the free product of* G_1 *and* G_2 *with amalgamated subgroup* H.

(II) *If* $(G_1, D_1, H_1, \Delta_1, S_1, B_1)$ *and* $(G_1, D_1, H_2, \Delta_2, S_2, B_2)$ *are large conglomerates,* $G \subset \mathcal{M}_n$ *is generated by* G_1 *and the cyclic group* $G_2 = \{\gamma\}$, *where* $\gamma \in \mathcal{M}_n, \gamma(S_1) = S_2, \gamma H_1 \gamma^{-1} = H_2$ *and* $\gamma(B_1) \cap B_2 = \varnothing$, *and if* $D = D_1 \backslash (B_1 \cup B_2 \cup S_2)$ *has nonempty interior and* $g(B_1 \cup S_1) \cap (B_2 \cup S_2) = \varnothing$ *for all* $g \in G_1$, *then* G *is Kleinian with fundamental set* D, *and each relation in it is a consequence of relations in* G_1 *and the relation* $\gamma H_1 \gamma^{-1} = H_2$.

30. Prove that if $G \subset \mathcal{M}_n$ is a geometrically finite Kleinian group, then all its subgroups of finite index are also geometrically finite.

31. Prove that every geometrically finite (nonelementary) Kleinian group $G \subset \mathcal{M}_n$ contains a subgroup that is not geometrically finite.

32. Prove that the stabilizer of any component of a geometrically finite Kleinian group in the plane is also geometrically finite.

33. Prove that b-groups cannot contain free abelian subgroups of rank 2.

34. Let $G \subset \mathcal{M}_n$ be a nonelementary Kleinian group. Then the set of points of approximation of it has Hausdorff dimensional greater than $n/2$.

35. Prove that a point of approximation of a Kleinian group $G \subset \mathcal{M}_n$ cannot lie on the boundary of a convex fundamental polyhedron P of G.

36. Prove that a Fuchsian group G is finitely generated if and only if $\Lambda(G)$ consists solely of points of approximation and parabolic fixed points.

37. Let $G \subset \mathcal{M}_n$ be a Kleinian group such that $\infty \in \Omega(G)$. Prove that for any sequence $\{g_k\}$ of distinct elements of G the radii r_{g_k} of their isometric spheres tend to zero and, moreover, the series

$$\sum_{g \in G \backslash G_\infty} r_g^q \quad \text{and} \quad \sum_{g \in G} |g'(x)|^q \quad (x \in \Omega(G))$$

converge for $q \geq 2n$, with the second series converging uniformly on compact subsets of $\Omega(G)$.

38. Two subgroups Γ and $\tilde{\Gamma}$ of a group G are said to be *commensurable* if the indices of $\Gamma \cap \tilde{\Gamma}$ in the groups Γ and $\tilde{\Gamma}$ are finite. Denote by $B(\Gamma) \subset G$ the group generated by all subgroups of G which are commensurable with the group Γ, and by $C(\Gamma)$ the group $\{g \in G \colon g\Gamma g^{-1}$ is commensurable with $\Gamma\}$.

Show that $B(\Gamma)$ is a normal subgroup of $C(\Gamma)$.

39. Show that if $G = SL(2, \mathbf{R})$ and Γ is the classical modular group (see Example 15), then in the preceding notation $B(\Gamma) = C(\Gamma) \supset SL(2, \mathbf{Q})$ and, consequently, these groups are nondiscrete.

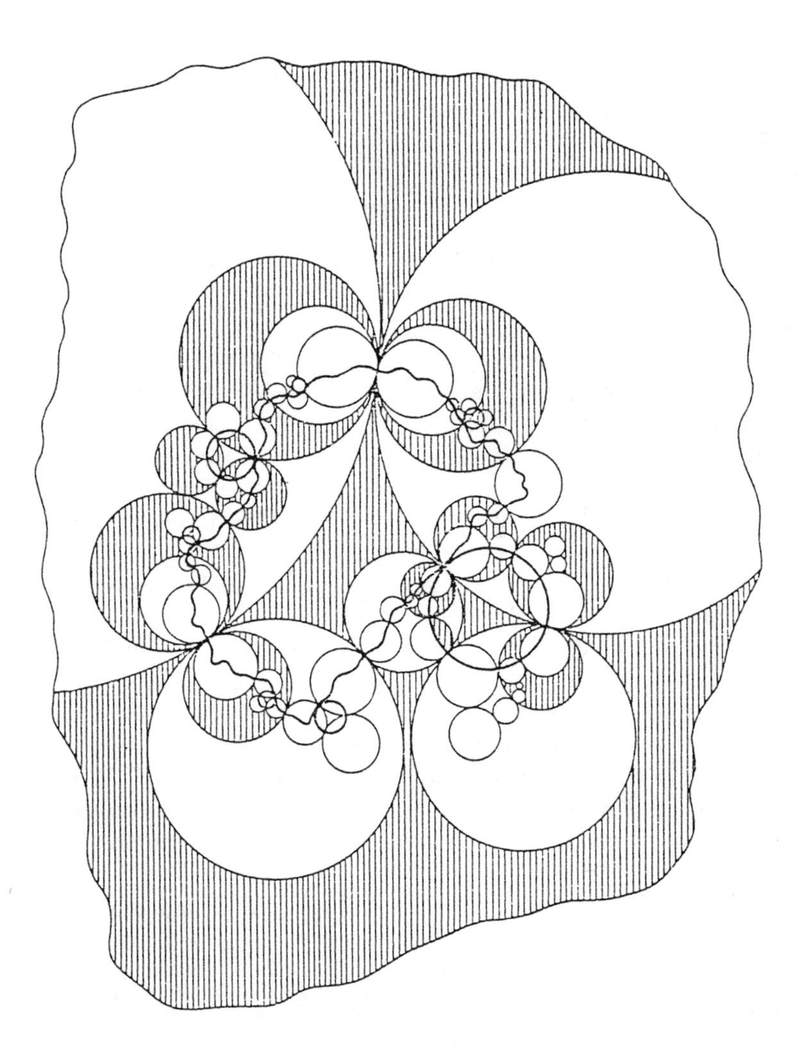

FIGURE 65

40. Find all the Kleinian group whose factor subgroups all coincide with the groups themselves.

41. Describe the factor subgroups of elementary groups.

42. Prove that all the factor subgroup of a Schottky group are trivial, and conversely.

43. Describe the factor subgroups of *b*-groups (see Figure 65).

44. Prove that every factor subgroup is elementary or quasi-Fuchsian or degenerate.

45. Prove that every Schottky group in $\overline{\mathbf{R}}^n$ is a quasi conformal deformation of a Fuchsian group of the second kind.

46. Let $\varphi\colon G \to G^*$ be a type-preserving isomorphism between finitely generated Fuchsian groups. Then G and G^* are groups of the same kind (first or second).

47. Let G be a finitely generated Kleinian group with invariant component. If each factor subgroup of G is cyclic, then there exists a type-preserving isomorphism of G onto a Fuchsian group of the second kind.

48. Let G_1 and G_2 be finitely generated Kleinian groups with invariant components Δ_1 and Δ_2, and let $\varphi\colon \Delta_1 \to \Delta_2$ be a similarity. Prove that φ^* preserves factor subgroups, i.e., if $H \subset G_1$ is a factor subgroup, then so is $\varphi^*(H) \subset G_2$.

49. Let G be a finitely generated torsion-free Kleinian group, and let G_0 be a finitely generated subgroup of it. Prove that if each element of G_0 leaves invariant at least $k \geq 1$ components of G, then G_0 itself leaves invariant at least k components of G.

50. Construct a doubly generated web group that is not Fuchsian and uniformizes two thrice-punctured spheres.

51. Construct a doubly generated *b*-group that uniformizes a punctured torus and a thrice-punctured sphere.

52. Construct doubly generated Kleinian groups with simply connected components that uniformize: a) a sphere with four points deleted; and b) a torus with two points deleted. Prove that if these groups are torsion-free, then they are $\mathbf{Z}_2$-extensions of quasi-Fuchsian groups.

53. Show that the Kleinian groups arising in Exercises 50–52 lie on the boundary of the classical Schottky space of genus 2 (see Problems 25 and 80).

§2. Exercises on the theory of uniformization

54. Show that the Riemann surface corresponding to the folium of Déscartes $x^3 + y^3 - 3xy = 0$ is conformally equivalent to a sphere.

55. Prove that a sphere with $n \geq 3$ points deleted is a surface of hyperbolic type, i.e., can be uniformized by a Fuchsian group.

56. Prove that there does not exist a quasi-Fuchsian group (that is not Fuchsian) which uniformizes a sphere with three points deleted.

57. Show that any triangulable orientable surface admits a conformal structure, and hence can be turned into a Riemann surface.

58. Prove that a closed Riemann surface with trivial one-dimensional homology group is conformally equivalent to a sphere.

59. Prove the ***Riemann-Hurwitz relation***: *If S and $\tilde{S}$ are closed Riemann surfaces of genera p and $\tilde{p}$, respectively, and $f\colon \tilde{S} \to S$ is an n-sheeted holomorphic covering, then*

$$2(\tilde{p}-1) = 2n(p-1) + \gamma, \tag{108}$$

where $\gamma = \sum_{x \in S}(\nu(x) - 1), \nu(x)$ being the branching order of f at a point $x \in S$.

60. Prove that the first-order holomorphic differentials $\varphi(z)\,dz$ on a closed Riemann surface S of genus p form a complex vector space $\Gamma_a(S)$ of dimension p.

61. Show that any (nonconstant) meromorphic function f on a closed Riemann surface S determines S as a covering of a sphere with the number of sheets equal to the order of f (the number of zeros or poles).

62. Prove that a sphere has a branched covering by any closed orientable surface. Give analytic and topological proofs.

63. Let $a_1, b_1, \dots, a_p, b_p$ be a basis in the one-dimensional homology group $H_1(S, \mathbf{Z})$ of a closed Riemann surface S of genus $p \geq 1$. Choose a basis $\varphi_1, \dots, \varphi_p$ in $\Gamma_a(S)$ such that $(\varphi_j, a_k) = \int_{a_k} \varphi_j = \delta_{jk}$, and let $(\varphi_j, b_k) = \int_{b_k} \varphi_j = \omega_{jk}$. Prove the following assertions:

1) The matrix $\Omega = (\omega_{jk})$ is nonsingular and symmetric, and $\operatorname{Im}\Omega$ is positive-definite.

2) Two surfaces S and S' of the same genus p and sufficiently close in the Teichmüller metric are conformally equivalent if and only if their period matrices Ω and Ω' coincide.

64. Show that a Riemann surface on which there are nonconstant bounded holomorphic functions is hyperbolic.

65. Show that if S is an open simply connected Riemann surface of hyperbolic type having a *Green's function*

$$G(z, z_0) = \ln(1/|z - z_0|) + g(z, z_0),$$

then a conformal mapping $f(z; z_0)$ of S onto the disk U with normalization $f(z_0; z_0) = 0$ is given by

$$f(z; z_0) = e^{-G(z,z_0) - iH(z,z_0)}, \tag{109}$$

where H is the harmonic conjugate function of G.

66. Construct a Riemann surface having a Green's function but not admitting nontrivial bounded harmonic functions.

67. Prove that if $z\colon S \to \overline{\mathbf{C}}$ is a meromorphic function on a compact Riemann surface S taking each value n times, then the following are true:

1) Every other meromorphic function f on S is connected with z by an nth-degree algebraic equation

$$f^n + r_1(z)f^{n-1} + \cdots + r_{n-1}(z)f + r_n(z) = 0, \tag{110}$$

where the $r_k(z), k = 1, \dots, n$, are rational functions of z.

2) f can be chosen so that equation (110), expressed in terms of z, is irreducible and so determines an algebraic function $f(z)$, and its Riemann surface is conformally equivalent to S.

68. Prove that if a conformal automorphism of a closed Riemann surface S is homotopic to the identity, then it is the identity mapping.

69. Prove the following ***theorem of Hurwitz***: *Let S be a closed surface of genus $p \geq 2$ with canonical dissection $a_1, b_1, \dots, a_p, b_p$ and let h be a conformal automorphism of S such that the loop $h(b_k)$ is homologous to b_k, $k = 1, \dots, p$. Then h is the identity automorphism.*

70. Prove that a conformal automorphism of a closed Riemann surface of genus p different from the identity can have at most $2p + 2$ fixed points.

71. Prove the following ***theorem of Hurwitz***: *The group* $\operatorname{Aut} S$ *of conformal automorphisms of a closed Riemann surface S of genus $p \geq 2$ has order $\leq$ $84(p-1)$.*

This estimate is sharp, since Klein constructed an example of a closed surface of genus 3 admitting 168 distinct conformal automorphisms (see also Problems 55–58).

72. Let S and S' be compact Riemann surfaces, and $H(S, S')$ the set of nonconstant holomorphic mappings of S onto S'. Prove that if the genus of S' is greater than 1, then $H(S, S')$ is finite (consider relation (108)).

73. Let $\pi'\colon S' \to S$ and $\pi''\colon S'' \to S$ be two coverings of a Riemann surface S, and $h\colon S' \to S''$ a homeomorphism of S' onto S'' such that $\pi' = \pi'' \circ h$. Prove that h is a conformal mapping.

74. Let $G \subset \mathcal{M}_2$ be a finitely generated function group with invariant component Δ. Show that if $w(z) \not\equiv \text{const}$ is a meromorphic function automorphic with respect to G (and having definite finite or infinite values at parabolic vertices of a fundamental region of G if such vertices exist), then the inverse function $z(w)$ is representable as a ratio of two solutions of the second-order linear differential equation $\eta''(w) = \varphi(w)\eta(w)$, where $\varphi(w)$ is an algebraic function of w; moreover, if $w(z)$ has simple poles, then $\varphi(w)$ is rational.

75. Describe the tangent and cotangent bundles over the Teichmüller space $T(p, n)$ (cf. Problem 72).

76. Show that $\operatorname{Mod} T(\Gamma)$ is a subgroup of $\operatorname{Mod} T(\mathbf{1})$ that acts invariantly on $T(\Gamma)$.

77. Prove that a pretzel (an orientable surface of genus 2) can be covered by any closed surface of greater genus, and this covering is finite-sheeted.

78. Describe all the three-sheeted coverings over a pretzel.

79. Prove that the fundamental group of a noncompact surface is always free.

80. Prove that every closed nonorientable surface S is a sphere with a definite number k of Möbius bands glued into it (k is called the *genus* of this surface), and that its fundamental group has the presentation

$$\pi_1(S) = \left\{ a_1, \dots, a_k \colon \prod_{j=1}^{k} a_j^2 = 1 \right\}. \tag{111}$$

81. Compute the fundamental groups of the projective plane and of the Klein bottle.

82. Prove that a closed nonorientable surface of genus k is homeomorphic to some orientable surface with a single Möbius band glued into it.

83. Show that every nonorientable surface has a two-sheeted orientable covering.

84. Prove that every nonorientable surface S of genus $k \geq 3$ (see (111)) can be uniformized by a discrete group G in the disk U that consists of conformal and anticonformal automorphisms of U, i.e., S is conformally or anticonformally equivalent to U/G.

85. Prove that if every loop on a triangulable surface S separates it, then S is orientable and homeomorphic to a region on a sphere.

In Exercises 86–102 it is assumed that all the Kleinian groups under consideration are finitely generated and do not contain elliptic elements.

86. Let G be a function group with invariant component Ω_0. Consider the surface $S_0 = \Omega_0/G$. Prove that the homomorphism $\pi_1(S_0) \to \pi_1(M(G))$ (see §7 in Chapter II) induced by the imbedding $S_0 \subset M(G)$ is an epimorphism.

87. Show that if Ω_0 is simply connected in the conditions of the preceding exercise, then the homomorphism $\pi_1(S_0) \to \pi_1(M(G))$ is an isomorphism.

88. Using the properties of the manifold $M(G)$, show that if G is a function group with invariant component Ω_0, then all the components of $\Omega(G) \backslash \Omega_0$ are simply connected.

89. Suppose that all the components of G are simply connected. Show that for any component S of the boundary of $M(G)$ the homomorphism $\pi_1(S) \to \pi_1(M(G))$ is a monomorphism, i.e., any (homotopically) nontrivial loop on S is nontrivial also in $M(G)$.

90. Let G be a Kleinian group such that the boundary of $M(G)$ is a torus. Show that G is then either a cyclic loxodromic group or a free abelian group of rank 2.

91. Suppose that the manifold M is homeomorphic to the product $T \times [0, 1]$, where T is a torus. Prove that there does not exist a Kleinian group G such that M is homeomorphic to $M(G)$, i.e., M cannot be uniformized by a Kleinian group.

92. Suppose that the boundary S of $M(G)$ is connected, and the homomorphism $\pi_1(S) \to \pi_1(M(G))$ is an isomorphism. Prove that $M(G)$ is not compact.

93. Let G_0 be a subgroup of a Kleinian group G of finite index such that $M(G_0)$ is homeomorphic to $S \times [0,1]$, where S is a Riemann surface of genus $p \geq 2$. Prove that G is then either a quasi-Fuchsian group or a $\mathbf{Z}_2$-extension of a quasi-Fuchsian group.

94. Suppose that the Kleinian group is such that $\partial M(G)$ is a closed surface of genus $g \geq 1$. Prove that if G is a free group of rank g, then G is a Schottky group.

95. Suppose that the manifold $M(G)$ is compact, where G is a Kleinian group. If the subgroup G_0 of G is finitely generated and such that $\Lambda(G_0) = \Lambda(G)$, then the index of G_0 in G is finite (cf. Exercise 19).([16])

96. Suppose that G is a Kleinian group isomorphic to the fundamental group of a closed surface that is orientable or nonorientable. Prove that if G has an invariant component, then this component is simply connected.

97. Suppose that G is a Kleinian group isomorphic to the fundamental group of a closed nonorientable surface. Prove that G does not have invariant components.

98. Suppose that G is a Kleinian group with connected set $\Omega(G)$ of discontinuity. If $\Omega(G)$ is not simply connected, then there is a simple loop α on the surface $S = \Omega(G)/G$ that is nontrivial on S and bounds a topological disk D in the manifold $M(G)$. Prove that if α separates S, then D separates $M(G)$.

99. Prove that if $M(G)$ is a handlebody of genus $p \geq 1$, then G is a Schottky group.

100. Let G be a Kleinian group with connected set of discontinuity. Prove that if $M(G)$ is compact, then G is a Schottky group.

101. Let G be a function group without parabolic elements. Prove that if G is free, then G is a Schottky group.

102. From the preceding exercises deduce the following ***theorem of Maskit***: *A Kleinian group G is a Schottky group if and only if G is free and does not contain parabolic elements.*

103. Let G be a subgroup of the additive group of rational numbers. Construct a three-dimensional manifold M such that $\pi_1(M)$ is isomorphic to G.

104. Let M be a compact three-dimensional manifold whose fundamental group $\pi_1(M)$ is torsion-free. Show that if $\pi_1(M)$ contains a nontrivial free subgroup of finite index, then $\pi_1(M)$ is itself free.

([16]) *Translation editor's note*: The statement is in general not true; for example, let G be a purely hyperbolic finitely generated Fuchsian group of the first kind, and let G_0 be its commutator subgroup.

105. Show that if the projective plane $\mathbf{RP}_2$ is imbedded in a three-dimensional manifold M, then the homomorphisms $i_*\colon H_1(\mathbf{RP}^2, \mathbf{Z}_2) \to H_1(M, \mathbf{Z}_2)$ and $i_*\colon \pi_1(\mathbf{RP}^2) \to \pi_1(M)$ are monomorphisms.

106. Prove that if S is a compact orientable surface with nonzero boundary ∂S, then $S \times I$ is a handlebody of genus $p = 1 - \chi(S)$, where $\chi(S)$ is the Euler characteristic of S.

107. Prove that: a) $L(1, q) = S^3$ for all $q = 1, 2, \ldots$, and b) if $q \equiv \pm q' \pmod{p}$ or $qq' \equiv \pm 1 \pmod{p}$, then $L(p, q)$ is homeomorphic to $L(p, q')$.

108. Prove that $P \# P = P \# (S^2 \times S^1)$, where P is a nonorientable 2-sphere bundle over the circle.

109. Prove that $L(p, q)$ is prime for all p and q.

110. Prove that all the 2-sphere bundles over the circle are prime.

111. Show that if p is relatively prime with q and q', then the manifolds $L(p, q)$ and $L(p, q')$ are homotopy equivalent (cf. (113)).

112. Suppose that M is a compact three-dimensional manifold with nonempty boundary ∂M. Prove that if ∂M contains a surface of positive genus, then $\pi_1(M)$ contain a subgroup of index 2; consequently, M has a two-sheeted covering manifold.

113. Prove that $L(2k, q)$ contains a closed surface S with Euler characteristic $\chi(S) = 2 - k$.

114. Let M be a three-dimensional compact manifold with nonempty boundary ∂M. Prove that if ∂M contains a surface of positive genus, then $H_1(M, \mathbf{Z})$ is infinite.

115. Let M be a three-dimensional compact irreducible manifold with infinite fundamental group $\pi_1(M)$. Prove that if $\pi_1(M)$ is a nontrivial direct product, then M is homeomorphic to $S \times S^1$, where S is a compact surface; moreover, if S is orientable, then $\pi_1(M)$ has the presentation

$$\begin{aligned} \pi_1(M) = \{a_1, b_1, \ldots, a_p, b_p, t\colon [a_i, t] = [b_i, t] = 1 \\ (i = 1, \ldots, p), [a_1, b_1] \cdots [a_p, b_p] = 1\}. \end{aligned} \tag{112}$$

116. Let M be a three-dimensional closed orientable manifold, and S a closed orientable surface of genus p. Prove that if $\pi_1(M)$ is isomorphic to $\pi_1(S) \times \mathbf{Z}$, then any Heegaard splitting of M has genus greater than $2p + 1$.

117. Let M be a three-dimensional closed orientable manifold, and S a closed orientable surface of genus p. Prove that if $\pi_1(M)$ is isomorphic to $\pi_1(S) \times \mathbf{Z}$ and M has a Heegard splitting of genus $2p + 2$, then M is homeomorphic to $S \times S^1$.

118. Prove that if a three-dimensional closed manifold M has a Heegaard splitting of genus 1 and $\pi_1(M)$ is isomorphic to $\mathbf{Z}$, then M is homeomorphic to the product of a circle and a sphere.

119. Let G be a given finitely generated group with finitely many defining relations. Prove that there exists a four-dimensional compact manifold whose fundamental group is isomorphic to G.

120. Show that for a lens space

$$\pi_1(L(p,q)) \cong H_1(L(p,q)) \cong \mathbf{Z}_p. \tag{113}$$

121. Prove that a closed simply connected three-dimensional manifold M is aspherical in all dimension (i.e., $\pi_n(M) = 0$ for all $n \geq 2$).

122. Show that if a manifold M is the Cartesian product of manifolds M_1 and M_2 ($M = M_1 \times M_2$), then $\pi_1(M)$ is the direct product of $\pi_1(M_1)$ and $\pi_1(M_2)$.

123. Compute the fundamental group of the n-dimensional torus.

124. Show that a spherical dodecahedron space has the homology type of a sphere.

125. Describe all the compact three-dimensional manifolds with boundary having free abelian fundamental groups.

126. Prove that the fundamental group of a complete hyperbolic manifold is torsion-free.

127. Prove that every abelian subgroup of the fundamental group of a compact hyperbolic manifold is cyclic.

128. Show that the fundamental group of an octahedron space is given by the relations $abc = adb = acd = bdc = \mathbf{1}$.

129. Show that the homology groups of an octahedron space have the form $H_0 = \mathbf{Z}, H_1 = \mathbf{Z}_3, H_2 = \mathbf{O}$ and $H_3 = \mathbf{Z}$.

130. Find the fundamental group of a hyperbolic dodecahedron space.

131. Let $T = \overline{U} \times [0,1]$ be a solid three-dimensional cylinder. Identify its bases $\overline{U} \times \{0\}$ and $\overline{U} \times \{1\}$ by means of an orientation-reversing homeomorphism. The (nonorientable) manifold thus obtained is called a *solid Klein bottle*. Prove that its fundamental group is isomorphic to $\mathbf{Z}$, and the solid torus serves as a two-sheeted orientable covering.

132. Describe all possible coverings of n-dimensional real projective space ($n \geq 2$).

133. Show that there are exactly two nonequivalent five-sheeted coverings of the manifold $S^3 \backslash L$ which is the complement of the *trifolium*, i.e., the knot with fundamental group $\pi_1(S^3 \backslash L) = \{x, y \colon xyx = yxy\}$.

134. Prove that all the Heegaard splittings of genus 1 (i.e., manifolds obtained by gluing together two solid tori along their boundaries) are lens spaces or are homeomorphic to $S^2 \otimes S^1$ (the converse is also true).

135. Prove that a complete n-dimensional Riemannian manifold of zero (sectional) curvature can be uniformized by a discontinuous group of Euclidean isometries of $\mathbf{R}^n$.

136. Prove that a compact Riemannian manifold of zero curvature can be covered by a torus, with the covering finite-sheeted.

137. Prove that the fundamental group of a complete hyperbolic manifold has trivial center and cannot be decomposed into a nontrivial direct product of subgroups.

§3. Problems: General properties of discontinuous groups

1. Characterize all the connected Lie subgroups of the Möbius group $\mathcal{M}_n$ ($n \geq 2$) (Greenberg [86]).

2. Let G be the real Möbius group $SL(2, \mathbf{R})/\{\pm 1\}$. Consider Fuchsian groups $\Gamma \subset G$ such that $\operatorname{meas}(G/\Gamma) < \infty$ (which is equivalent to the non-Euclidean area of a fundamental region of Γ being finite and, consequently, to Γ being finitely generated),(17) and introduce the following topology in the set of conjugacy classes $[\Gamma]$ of Fuchsian subgroups of G: Take a system of generators $\gamma_1, \dots, \gamma_r$ of a group $\Gamma \subset [\Gamma]$ and a neighborhood v of the identity in G; then a neighborhood V of the class $[\Gamma]$ consists of the conjugacy classes $[\Gamma']$ for which there exists an isomorphism $\chi\colon \Gamma \overset{\text{onto}}{\longrightarrow} \Gamma' \in [\Gamma']$ such that $\chi(\gamma_j) \circ \gamma_j^{-1} \in v$ for all $j = 1, \dots, r$, and, moreover, it is also required that χ and χ^{-1} preserve the type of parabolic elements.

Prove the following assertion: *The set of conjugacy classes* $[\Gamma]$ *for a Fuchsian group* Γ *such that* $\operatorname{meas}(G/\Gamma) \leq \mu < \infty$ *and the trace of each hyperbolic element* $\gamma \in \Gamma$ *satisfies the inequality* $|\operatorname{tr}\gamma| \geq 2 + \varepsilon, \varepsilon > 0$, *is compact* (Mumford [195], Bers [53]).

3. Let P be a compact convex polyhedron in an n-dimensional simply connected complete space X of constant curvature, and for each $(n-1)$-dimensional face A of P let g_A be an isometry of X such that $P \cap g_A(P) = A$. Suppose, moreover, that 1) for each $(n-1)$-dimensional face A of P there exists a face A' such that $g_A g_{A'} = 1$, and 2) for each $(n-2)$-dimensional face E of P there exists a sequence $A_1, \dots, A_k$ of $(n-1)$-dimensional faces such that

$$g_{A_1} g_{A_2} \cdots g_{A_k} = 1, \qquad P \cap g_{A_1}(P) \cap g_{A_1} g_{A_2}(P) \cap \cdots \cap g_{A_1} \cdots g_{A_{k-1}}(P) = E$$

and the polyhedra $P, g_{A_1}(P), \dots, g_{A_1} \cdots g_{A_{k-1}}(P)$ do not have pairwise common interior points.

Then the group G of isometries of X generated by the transformations g_A is discrete, and P is a fundamental region of it.

This follows from a more general result of Aleksandrov [25] on filling a space by convex polyhedra; see also [1].

The condition that P be compact is essential in the case of the hyperbolic space H^n; see [234].

This result (as well as the results in Problems 7 and 9) remains valid also for unbounded polyhedra if in addition we have the following local finiteness condition for the tesselation: any compact subset of H^n intersects only finitely many images of the polyhedron P. This condition is equivalent to the condition that the boundary $\partial P \cap \partial H^n$ of P does not contain fixed points of the isometries g_A (see above) that are hyperbolic translations (perhaps helical);

(17) *Translation editor's note*: The last equivalence only holds for groups of the first kind; for a group of the second kind, the area of a fundamental region is always infinite.

see [183], [257] and [261]. For $n \geq 4$ the noncompact case is similar to the compact case, i.e., deformations of tesselations of the hyperbolic space H^n, $n \geq 4$, are strongly rigid in the sense that if the assertion of this part (and of Problem 7) is valid for a polyhedron P, then it remains in force (without the additional condition that the tesselation be locally finite) also for all polyhedra $P' \subset H^n$ of the same combinatorial type as P; this result of Apanasov can be obtained for small deformations from a result of Garland and Raghunathan [81] on filling a whole neighborhood of a discrete group $G \subset \operatorname{Isom} H^n$, $n \geq 4$ (in the topology of pointwise convergence of representations) by groups conjugate to it.

4. Prove that a group $G \subset \mathcal{M}_2$ is Kleinian if and only if each doubly generated subgroup of it is Kleinian (Jørgensen [109]).

5. Prove that a group $G \subset \mathcal{M}_2$ which leaves invariant the disk is a discrete (Fuchsian) group if and only if every cyclic subgroup of it is discrete (Jørgensen [109]).

6. Prove that if a discrete group $G \subset \mathcal{M}_2$ with generators X and Y is nonelementary, then

$$|\operatorname{tr}^2(X) - 4| + |\operatorname{tr}(XYX^{-1}Y^{-1}) - 2| \geq 1 \tag{114}$$

(Jørgensen [109]).

7. Prove that the n-dimensional hyperbolic space $\mathbf{R}^n_+$ can be filled by reproducing a regular compact polyhedron $P \subset \mathbf{R}^n_+$ along full faces if and only if all its dihedral angles constitute 2π divided by an integer (Aleksandrov [25]).

8. Let P be a convex polyhedron (in $\mathbf{R}^n$ or in the hyperbolic space $\mathbf{R}^n_+$). Prove that the group G generated by the reflections in the $(n-1)$-dimensional faces is discrete if and only if all the dihedral angles of P are equal to 2π divided by an integer, and if such an angle at an $(n-2)$-dimensional edge $Q \subset P$ is 2π divided by an odd integer, then P must have a plane of symmetry passing through Q (Aleksandrov [25]).

9. Let P be a bounded non-Euclidean polygon in the unit disk U bounded by an even number of sides $s_1, s_1', \ldots, s_n, s_n'$. Assume that there exist conformal automorphisms $\gamma_1, \ldots, \gamma_n$ of U such that $\gamma_1(s_1) = s_1', \ldots, \gamma_n(s_n) = s_n'$. Let z_1 be a vertex of P. A *cycle* of z_1 is defined to be a set of vertices of P equivalent to it with respect to the automorphisms $\gamma_1, \ldots, \gamma_n$. Let $\{z_1, \ldots, z_m\}$ be an arbitrary cycle, with $\gamma_{i_1}(z_1) = z_2, \gamma_{i_2}(z_2) = z_3, \ldots, \gamma_{i_m}(z_m) = z_1, g_m = \gamma_{i_m} \circ \cdots \circ \gamma_{i_1}$, and let $\alpha(z_i)$ be the angle formed by the sides of P intersecting at z_i. Prove the following ***theorem of Poincaré***:

Let Γ be the group generated by $\gamma_1, \ldots, \gamma_n$. If for every cycle $\{z_1, \ldots, z_m\}$ of vertices of P there exists an integer ν such that $\sum_1^m \alpha(z_j) = 2\pi$, then Γ is discontinuous (Fuchsian), and P is a fundamental polygon of it; moreover, each relation in Γ is a consequence of relations of the form $g_m^\nu = I$.

A similar assertion with appropriate modifications is true also for polygons with infinitely many sides, as well as for polyhedra in hyperbolic space (Poincaré [207], Maskit [183]).

10. For any discrete group Γ of isometries of H^n, any point $x \in H^n$, and any $\varepsilon > 0$, define the subgroup $\Gamma_\varepsilon(x)$ generated by the elements $\gamma \in \Gamma$ such that $d(x, \gamma(x)) \leq \varepsilon$, and let $\Gamma'_\varepsilon(x) \subset \Gamma_\varepsilon(x)$ be the subgroup consisting of the elements γ such that $d(x, \gamma(x)) = \varepsilon$.

Prove the ***lemma of Margulis*** [173]: *For each dimension n there exists an $\varepsilon > 0$ such that, for any discrete group Γ of isometries of H^n and for any point $x \in H^n$, the group $\Gamma'_\varepsilon(x)$ is abelian and $\Gamma_\varepsilon(x)$ has an abelian subgroup of finite index.*

11. Prove that if $G \subset \mathcal{M}_n$ is geometrically finite (i.e., has a fundamental polyhedron in $\mathbf{R}^{n+1}_+$ with finitely many faces), then each convex fundamental polyhedron in $\mathbf{R}^{n+1}_+$ of it has finitely many faces (Beardon and Maskit [44] for $n = 2$; Apanasov [253] for $n \geq 3$).

12. Let $G \subset \mathcal{M}_n$ be a Kleinian group in $\mathbf{R}^n_+$. Show that the set of $y \in \mathbf{R}^n_+$ such that the fundamental polyhedron $F_y(G) \cap \mathbf{R}^n_+$ has equivalent parabolic vertices on the boundary lies in the union of at most countably many $(n-1)$-dimensional spheres (Apanasov [34]).

13. Suppose that a Kleinian group G in $\mathbf{R}^n_+$ $(n \geq 3)$ is such that the fixed points of any elements g of it lie on the isometric sphere $I(g)$. Prove that its isometric fundamental polyhedron has finitely many faces (Apanasov [29]).

14. Prove the following assertions, which give conditions under which a given set $P \subset \overline{\mathbf{C}}$ is a fundamental set for a given Kleinian group G:

a) Let Δ_0 be a simply connected component of a finitely generated Kleinian group G without elliptic elements, and let D be a region in Δ_0 whose boundary consists of finitely many Jordan arcs which are pairwise equivalent with respect to the generators of the subgroup G_{Λ_0}; further, suppose that under the action of these generators (after identification of equivalent boundary points) D gives a Riemann surface homeomoprhic to Δ_0/G_{Δ_0}. Then D (after a part of the boundary points is adjoined to it) is a fundamental region for G_{Δ_0} in Δ_0.

b) Suppose that G_0 is a finitely generated subgroup of a Kleinian group G and is the stabilizer of a simply connected component $\Delta_0 \subset \Omega(G)$, and let P be a region in $\overline{\mathbf{C}}$ whose boundary consists of finitely many Jordan arcs that are pairwise equivalent with respect to generators of G_0; moreover, suppose that this region and its images under the action of these generators give a Riemann surface homeomorphic to Δ_0/G_{Δ_0} and such that the vertices of some fundamental polygon of G_0 in Δ_0 are interior points and are not branch points (in totality, the region P and the regions $g(P)$ $(g \in G_0)$ surrounding it are located in a one-to-one manner in $\overline{\mathbf{C}}$). Then P is a fundamental region of G_0 in Δ_0.

Roughly speaking, these assertions show that every region on the plane defining a Riemann surface homeomorphic to the original one can be taken as a fundamental region for G_0 (Krushkal′ [136], [137]).

15. Prove that for $n \geq 5$ there are no discrete groups G of hyperbolic motions of $\mathbf{R}^n_+$ such that a fundamental polyhedron $P(G)$ is a bounded simplex; for $n = 4$ there are five such groups, and for $n = 3$ there are nine (Vinberg [235]).

16. Prove the following:

a) For $n \geq 10$ there are no discrete groups of motions of the hyperbolic space H^n that are generated by reflections in the faces of a hyperbolic simplex (Lanner [284], Vinberg [235]).

b) For $n \geq 30$ there are in general no discrete groups of motions in the hyperbolic space H^n that are generated by reflections in the faces of a compact convex polyhedron, nor are there arithmetic discrete groups (of reflections) of noncompact type.

c) For $n \geq 22$ there are no arithmetic reflection groups in H^n of compact type with field of definition other than $\mathbf{Q}(\sqrt{5}), \mathbf{Q}(\sqrt{2})$, and $\mathbf{Q}(\cos(2\pi/7))$.

d) For $n \geq 14$ there are no arithmetic reflection groups in H^n of compact type with field of definition other than $\mathbf{Q}(\sqrt{2}), \mathbf{Q}(\sqrt{3}), \mathbf{Q}(\sqrt{5})$, $\mathbf{Q}(\sqrt{6})$, $\mathbf{Q}(\sqrt{2}, \sqrt{3})$, $\mathbf{Q}(\sqrt{2}, \sqrt{5})$, and $\mathbf{Q}(\cos(2\pi/m))$, where $m = 7, 9, 11, 15, 16$, or 20.

Assertions b)–d) were proved by Vinberg in [308] and [309] with the use of an estimate of Nikulin (see [295] and [296]) for the average complexity of the faces of a simple polyhedron.

Here an arithmetic discrete group of reflections (in the sense of the theory of discrete subgroups of semisimple Lie groups; see [40]) is understood to be a group with matrix group representation (see Chapter I, §2) giving a subgroup of finite index in the group of integral linear transformations preserving a suitable integral quadratic form of signature $(n, 1)$. Examples of such groups are known only for $n \leq 19$ (see [306], [307] and [310]).

17. Prove that if a Fuchsian group $G \subset \mathfrak{M}_n$ acting in the ball B^n is such that the measure of the set of points of approximation (see §2 in Chapter I) in $\Lambda(G)$ is greater than the measure of the set of residual limit points, then either $\Lambda(G) = \partial B^n$ or $\mu_{n-1}(\Lambda(G)) = 0$ (Chen Su-Shing [65]).

18. Let $G \subset \mathfrak{M}_n$ be a discontinuous group of Möbius mappings such that condition (63) holds and $\mu_{n-1}(\Lambda(G)) > 0$. For an arbitrary set $E \subset \Lambda(G)$ it is possible to define the multiplicity function

$$k(x; E) = \operatorname{card} E \cap Gx, \qquad x \in \Lambda(G). \tag{115}$$

Prove the following assertions:

a) For a Borel set E the multiplicity function (115) is a Borel function, and for a Lebesgue-measurable set E it is also L-measurable.

b) If E is a measurable subset of $\Lambda(G)$ of positive m_{n-1}-measure, then the function $k(x, E)$ is ∞ almost everywhere on E, i.e., the intersection $E \cap Gx$ either is empty or consists of infinitely many points (Krushkal′ [135]).

19. Prove that under the conditions of the preceding problem the equation $\varphi(g(x))|g'(x)|^p = \varphi(x)$ $(g \in G, x \in \Lambda(G))$ with $p \in \mathbf{Z}\setminus\{0\}$ has only the zero solution in the class of measurable functions that are almost everywhere finite on $\Lambda(G)$.

In particular, every measurable G-form $\varphi(z)\, dz^p\, d\bar{z}^q$ $(z \in \mathbf{C})$ concentrated on $\Lambda(G)$ can take only the values 0 or ∞ when $p + q \neq 0$ (Krushkal′ [135]).

20. Let G be a Kleinian group of Möbius automorphisms of $\mathbf{R}^n_+$ and $\Lambda(G) = \overline{\mathbf{R}^{n-1}} = \partial\mathbf{R}^n_+$. Prove that the following conditions are equivalent:

1) $m_{n-1}(\partial F(G) \cap \Lambda(G)) = 0$.

2) Almost all points of $\Lambda(G)$ are points of approximation.

3) If a function $u(x, y)$ is bounded and harmonic in each variable $x \in \mathbf{R}^n_+$ and $y \in \mathbf{R}^n_+$, and $u(g(x), g(y)) = u(x, y)$ for all $g \in G$, then u is constant, i.e., the action of G along geodesics in $\mathbf{R}^n_+$ is ergodic.

4) Each G-invariant bounded harmonic function $u(x)$ on $\mathbf{R}^n_+$ has nontangential limits $\lim u(x)$ as $x \to y$ for almost all $y \in \partial\mathbf{R}^n_+$ (Hopf; Sullivan [228]).

Then $u(x) = \text{const}$ by the maximum principle; it follows from this, in particular, that the action of G on $\partial\mathbf{R}^n_+$ is ergodic.

21. Prove that Maskit-Klein combination of groups with limit sets of zero measure again gives a group with the same property; in particular, all nondegenerate b-groups are of this kind (Maskit [181]).

22. Let G_1 and G_2 be geometrically finite Kleinian groups. Prove that a Kleinian group G obtained from G_1 and G_2 by Maskit combination is also geometrically finite (Marden [168]).

23. Prove that a geometrically finite torsion-free Kleinian group with connected set of discontinuity can be constructed from elementary groups by Klein combination (Gusevskiĭ[92], [94]).

24. Prove that any finitely generated torsion-free Kleinian group with connected set of discontinuity can be constructed from finitely many elementary and degenerate groups by Klein combination (Gusevskiĭ [92], [94]).

25. Call a Schottky group *classical* if it is generated by some system of mappings whose defining curves are circles (cf. §1 in Chapter I).

Prove that there exist Schottky groups that are not classical (Marden [170]).(18)

26. Let G_i be the stabilizer of a component Ω_i of a Kleinian group G such that Ω_i/G_i is a Riemann surface of finite type $(i = 1, 2)$. Prove that $\Lambda(G_1 \cap G_2) = \Lambda(G_1) \cap \Lambda(G_2) = \overline{\Omega}_1 \cap \overline{\Omega}_2$ (Maskit [185]).

27. Prove that the limit set of a Kleinian group $G \subset \mathcal{M}_3$ cannot be a tame knot (Kulkarni [145]).

---

(18) *Translation editor's note*: See also Zarrow [249].

28. Prove that if a Kleinian group G with N generators contains only loxodromic (hyperbolic) elements, then the number of components of $\Omega(G)/G$ does not exceed $N/2$ (Marden [167]).

29. Prove that a b-group G is nondegenerate if and only if it is geometrically finite (Abikoff [6]).

30. Is a normal finite extension of a stable group quasiconformally stable (see Kra [123])?

31. A Kleinian group G is said to be *Maskit-constructible* if it is finitely generated and can be obtained from cyclic groups by combination. Prove that each Maskit-constructible group is uniformly stable, i.e., all its quasiconformal images are also stable (Abikoff [5]).

32. Prove that each nondegenerate b-group is a boundary group (Il′yashenko [105], [106]).

33. Prove that if Ω_0 is an invariant component of a nondegenerate b-group G, then a conformal mapping $f\colon U \to \Omega_0$ admits a continuous extension to the closed disk $\overline{U}$ (Abikoff [6]).

For the remaining components $\Omega_j \subset \Omega(G)\backslash\Omega_0$ this is obvious, since they are bounded by quasicircles.

34. Prove that a Kleinian group $G \subset \mathcal{M}_n$ is Fuchsian if it consists of hyperbolic mappings, parabolic mappings without torsion, and elliptic mappings of genus $n-2$ (Apanasov [29]).

35. Suppose that a sequence of Kleinian groups $G_m \subset \mathcal{M}_n$ with generators $T_{1,m}, \dots, T_{k,m}, 2 \leq k < \infty$, is such that

1) the $T_{i,m}$ are nonelliptic for all i and m,

2) $T_{i,m}$ and $T_{j,m}$ do not have common fixed points for any m and $i \neq j$, and

3) $\lim_{m\to\infty} T_{i,m} = T_i \in \mathcal{M}_n$ for all $i = 1, \dots, k$.

Prove that then $T_i \neq I$ for all $i = 1, \dots, k, 2 \leq k < \infty$ (Apanasov [30]).

36. Let $G(n) = \{T_j(n),\ j = 1, 2, \dots\}$ be a sequence of Kleinian groups $(n = 0, 1, \dots)$. Assume that there exist Möbius mappings T_j such that $T_j = \lim_{n\to\infty} T_j(n)$. Denote by G the group generated by $T_0, T_1, \dots$. Assume also that all the mappings $T_j(0) \to T_j(n)$ from $G(0)$ onto $G(n)$ are type-preserving isomorphisms. Then the mapping $T_j(0) \to T_j$ is an isomorphism of $G(0)$ onto G, and G does not contain elliptic elements of infinite order (Chuckrow [66]).

37. Let $G_0 \subset \mathcal{M}_2$ be a nonelementary discrete group, and let $\{\varphi_n\}$ be isomorphisms of it onto discrete groups G_n. Prove that the group G consisting of the Möbius transformations $\varphi(g) = \lim_{n\to\infty} \varphi_n(g)$, $g \in G_0$, is discrete, and φ is an isomorphism of G_0 onto G (Jørgensen [109]).

38. Prove that there exists an $r = r(n) > 0$ such that a ball of hyperbolic radius r can be inscribed in a fundamental polyhedron of any Fuchsian group G without elements of finite order in the ball B^n (Apanasov [30]).

For $n = 2$ and $n = 3$ this assertion was obtained for arbitrary Fuchsian groups (see [169], [226] and [245]).

39. Let G be a Kleinian group of finite type, and $\Sigma \subset \Omega(G)$ an invariant union of components of G. Prove that if f is a topological automorphism of Σ commuting with the elements of G, i.e., $f \circ g = g \circ f$ for all $g \in G$, then (i) f extends to a topological automorphism $\hat{f}$ of $\overline{\mathbf{C}}$ such that $\hat{f}|_{\overline{\mathbf{C}} \setminus \Sigma} = I$, and (ii) if f is quasiconformal, then so is $\hat{f}$ (an extension theorem of Maskit; see [182] and [55]).

40. Let $G_1, G_2 \subset \mathcal{M}_n, n \geq 3$, be Fuchsian groups in the ball B^n such that $\mathrm{vol}(B^n/G_i) < \infty$, $i = 1, 2$, and $G_1 = fG_2f^{-1}$ for some diffeomorphism $f\colon B^n \to B^n$. Prove that this isomorphism of G_1 and G_2 or of subgroups of finite index in them is an inner isomorphism in the group $\mathcal{M}_n$ (Garland and Raghunathan [81]).

For $n = 3$ this assertion can be obtained from results of Marden [168].

Let G be a finitely generated Kleinian group with invariant component Δ, and let $\pi\colon \Delta \to \Delta/G = S$ be the natural projection. Then S is a closed Riemann surface with n marked points $x_1, \dots, x_n$, each of which is assigned a branching order ν_i ($2 \leq \nu_i \leq \infty$). In §4 of Chapter II it was shown that there exist a finite set of simple disjoint loops $v_1, \dots, v_m$ on S, none passing through the points $x_1, \dots, x_n$, and integers α_i ($1 \leq \alpha_i \leq \infty$) such that $v_i^{\alpha_i}$ lifts to a loop in Δ when $\alpha_i < \infty$, while if $\alpha_i = \infty$, then the element of G defined by the loop v_i is parabolic.

Let us construct a two-dimensional complex K as follows: Cut the surface S along the loops v_i, and glue a disk to S along each of them. If $\alpha_i > 1$, then assign to the center of each disk the number α_i, and join these centers by a segment; and if $\alpha_i = 1$, then leave these disks unchanged. The resulting two-dimensional complex K consists of finitely many surfaces with marked points (with integers assigned to them) and of one-dimensional complexes joining two marked points with the same assigned numbers.

The pair (p, K), where p is the genus of the surface $S = \Delta/G$ and K is the complex constructed above, is called the *signature* of the group G under consideration.

Two Kleinian groups G_1 and G_2 with invariant components Δ_1 and Δ_2 are said to be *similar* if there exists an orientation-preserving homeomorphism $\varphi\colon \Delta_1 \to \Delta_2$ (called a *similarity*) such that the homomorphism $\varphi^*\colon G_1 \to G_2$, defined by the rule $\varphi^*(g) = \varphi \circ g \circ \varphi^{-1}$, is an isomorphism preserving the type of elements, i.e., φ^* and $(\varphi^*)^{-1}$ preserve parabolic elements, and φ^* preserves the square of the trace of each elliptic element.

A finitely generated Kleinian group G with invariant component is called a *Koebe group* if each factor subgroup of G is either elementary or Fuchsian (cf. §4 in Chapter II).

41. Prove that each finitely generated Kleinian group G with invariant component is conformally similar to some Koebe group G^* (Maskit [188]).

42. Prove that if there exists a type-preserving isomorphism of a Kleinian group G onto a finitely generated Fuchsian group of the first kind, then G

has a simply connected invariant component and does not contain accidental parabolic elements (Maskit [188]).

43. Prove that two finitely generated function groups G_1 and G_2 are similar if and only if they have the same signature (Maskit [188]).

44. Let G_1 and G_2 be geometrically finite function groups with the same signature. Prove that there exists a quasiconformal homeomorphism $\varphi\colon \Omega(G_1) \to \Omega(G_2)$ such that the homomorphism $\varphi^*\colon G_1 \to G_2$ defined by the rule $\varphi^*(g) = \varphi \circ g \circ \varphi^{-1}$ is an isomorphism that preserves the type of elements. Moreover, φ can be extended to a global quasiconformal homeomorphism $\hat{\varphi}\colon \overline{\mathbf{C}} \to \overline{\mathbf{C}}$ (Maskit [188]).

45. Let G and H be Kleinian groups satisfying the following conditions:

1) G is geometrically finite,

2) there exists an orientation-preserving homeomorphism $f\colon \Omega(G) \to \Omega(H)$ that induces an isomorphism $\varphi\colon G \to H$.

Prove that there then exists a quasiconformal homeomorphism $g\colon \overline{\mathbf{R}}^3_+ \to \overline{\mathbf{R}}^3_+$ of the closed half-space which induces φ. If f is a quasiconformal homeomorphism, then f has a quasiconformal extension to $\overline{\mathbf{C}}$. If f is conformal, then φ is an inner automorphism (Marden [168]).

An isomorphism $\varphi\colon G \to H$ between Kleinian groups G and H is said to be *geometric* if there exists a homeomorphism f of $\mathbf{R}^3_+ \cup \overline{\mathbf{C}}$ onto itself such that $f \circ \gamma(x) = \varphi(\gamma) \circ f(x)$ for all $\gamma \in G$ and for all $x \in \overline{\mathbf{R}}^3_+$. In this case f will be said to *realize* φ.

An element $\gamma \in G$ is called a *boundary* element if γ belongs to the stabilizer G_Δ of some component Δ of G. A boundary element $\gamma \in G$ is called an *a-boundary* element if it belongs to the stabilizer G_Δ of some component Δ of G, and G_Δ is not a maximal subgroup of G.

46. Let G be a geometrically finite torsion-free Kleinian group having more than two components, with each simply connected. Further, let $\varphi\colon G \to H$ be an isomorphism of Kleinian groups that satisfies the following conditions:

a) $\varphi(\gamma)$ is parabolic if and only if γ is,

b) $\varphi(\gamma)$ is a boundary element if and only if γ is,

c) $\varphi(\gamma)$ is an a-boundary element if and only if γ is.

Prove that φ is a geometric isomorphism (Marden and Maskit [171]).

47. Let G be a geometrically finite purely loxodromic Kleinian group having more than two components, each simply connected, and let H be a purely loxodromic Kleinian group. Prove that if $\varphi\colon G \to H$ is an isomorphism preserving boundary elements into boundary elements, then φ is a geometric isomorphism (Marden and Maskit [171]).

48. Prove that nonelementary doubly generated torsion-free Kleinian groups are free groups (Gusevskiĭ [94]).

49. Prove that the geometrically finite doubly generated torsion-free Kleinian groups are exhausted by the following groups: 1) groups of Schottky type; 2) Fuchsian groups that uniformize two thrice-punctured spheres; 3) web groups that have an infinite set of components and uniformize two thrice-punctured spheres; 4) quasi-Fuchsian groups which uniformize two punctured tori; 5) b-groups which uniformize a torus with a point deleted and a thrice-punctured sphere; 6) $\mathbf{Z}_2$-extensions of quasi-Fuchsian groups which uniformize a sphere with four points deleted; and 7) $\mathbf{Z}_2$-extensions of quasi-Fuchsian groups which uniformize a twice-punctured torus (Gusevskiĭ [94]).

50. Describe the three-dimensional manifolds $M(G)$ that can be uniformized by doubly generated geometrically finite torsion-free Kleinian groups. In particular, show that the manifolds $M(G)$ corresponding to groups of the types 6) and 7) in the preceding problem are nontrivial line bundles over noncompact nonorientable surfaces (find them) (Gusevskiĭ [94]).

51. Show that the doubly generated Fuchsian groups $G = \langle X, Y \rangle \subset \mathcal{M}_2$, for which equality is attained in (114) are exhausted by the triangular groups with signature $(2, 3, q)$, $q = 7, 8, 9$. At the same time, the class of such doubly generated Kleinian groups has the cardinality of the continuum (Jørgensen and Kiikka [111]).

52. Let $\Gamma \subset M_n$ be discrete, and let $\mathrm{vol}(\mathcal{M}_n/\Gamma) < \infty$. Show that in this case (in the notation of Exercise 38) either $C(\Gamma)$ is discrete and the index of Γ in $C(\Gamma)$ is finite, or the group $B(\Gamma)$ is dense in $\mathcal{M}_n$ (Greenberg [88]).

53. Let (again in the notation of Exercise 38) $G = SL(2, \mathbf{R})$, and suppose that $\Gamma = \Gamma_q$ is generated by the two mappings $f(z) = -1/z$ and $g(z) = z + 2\cos(\pi/q), q = 3, 4, \ldots$ (the qth Hecke group). Show that $C(\Gamma_q) = \Gamma_q$ for $q \geq 5$, $q \neq 6$, and, consequently, $B(\Gamma)$ and $C(\Gamma)$ are discrete (Greenberg [88]).

§4. Problems: Methods of uniformization

54. Prove that for discrete groups $\Gamma \subset \mathcal{M}_n$ with $\mathrm{vol}(\mathcal{M}_n/\Gamma) < \infty$ the group $C(\Gamma) = \{g \in \mathcal{U}_n : g\Gamma g^{-1}$ is commensurable with $\Gamma\}$ (see Exercise 38 and Problems 52 and 53) is dense in $\mathcal{M}_n$ if and only if Γ is an arithmetic group. A proof that $C(\Gamma)$ is dense in $\mathcal{M}_n$ for arithmetic groups Γ was given by Greenberg [88]. The converse assertion on the arithmeticity of Γ is due to Margulis [288] and is a consequence of the following theorem of Margulis on super-rigidity (cf. [289]):

Let Γ be a lattice in a connected semisimple Lie group G with trivial center, let k be a local field, let $m > 1$, and let $\rho\colon C(\Gamma) \to GL_m(k)$ be an m-dimensional irreducible representation of the group $C(\Gamma)$ over k. Assume also the following conditions:

1) *The Zariski closure $\overline{\rho(\Gamma)}$ of the group $\rho(\Gamma)$ is a connected k-simple group.*
2) *$\rho(\Gamma)$ is not relatively compact in $GL_m(k)$ in the k-topology.*

Then([19])

([19]) *Translation editor's note:* Statements a) and b) are literal translations of material submitted by the second author.

a) *k is* **R** *or* **C** *(in other words, if k is totally disconnected and condition* 1) *holds, then $\rho(\Gamma)$ is relatively compact in the k-topology);*

b) *if k is* **R** *or* **C**, *then ρ extends to a continuous representation of G.*

55. Prove that equality in the Hurwitz theorem on the number $N(p)$ of conformal automorphisms of a closed Riemann surface of genus p (see Exercise 71) is attained only for $p = 3$ when $2 \leq p \leq 6$ (Wiman [247]).

56. Prove that equality in the Hurwitz theorem $N(p) \leq 84(p-1)$ is attained for an infinite sequence of numbers p (Macbeath [158]).

57. Construct an algebraic curve (a Riemann surface) of genus 7 having a group of birational automorphisms with order $504 = 84(7-1)$; this group is isomorphic to $PSL(2, 2^3)$ (Macbeath [159]).

To within a birational equivalence the algebraic Klein curve of genus 3 in Exercise 71 has the form $x^3y + y^3z + z^3x = 0$ (in homogeneous coordinates).

58. Prove that $N(p) \geq 8(p+1), p \geq 2$, where this lower bound is sharp and is attained for infinitely many distinct values of p (Accola [11], Macbeath [160]).

59. Let S be a closed Riemann surface of genus $p > 1$, and let $f \in \operatorname{Aut} S$. Prove that if there exist four homologically independent loops $\alpha_1, \alpha_2, \alpha_3$, and α_4 such that the intersection indices have values $(\alpha_1, \alpha_3) = 1, (\alpha_2, \alpha_4) = 1$ and $(\alpha_i, \alpha_j) = 0$ for $i + j \equiv 1 \pmod 2$ and $f(\alpha_i)$ is homologous to α_i $(i = 1, 2, 3, 4)$, then the automorphism f is the identity (Accola [10]).

60. Let X and Y be closed Riemann surfaces, and $H(X, Y)$ the set of surjective holomorphic mappings of X onto Y. Prove that there exists a number $N_1(p)$, depending only on the genus p of X, such that $\operatorname{card} H(X, Y) \leq N_1(p)$ for any surface Y of genus greater than 1.

A similar result holds also for nonsingular projective manifolds of arbitrary dimension with ample canonical bundles, except that the number N_1 then depends on the characteristic Hilbert polynomial of the manifold X (Bandman [263]).

Prove that the number of surjective holomorphic mappings of a Riemann surface X of genus $p > 1$ onto a Riemann surface Y of genus greater than 1 does not exceed

$$(2^{2+p(p-1)/2}\sqrt{6}(p-1)+1)^{2+2p^2}p^2(p-1) + 84(p-1)$$

(Howard and Sommese [276]).

61. Prove that the group of classes of topological automorphisms (which is isomorphic to the quotient of $\operatorname{Aut} \pi_1(S)$ by the subgroup of the inner automorphisms of $\pi_1(S)$) of a closed Riemann surface S of genus p is generated by the windings with respect to the curves $\{a_i, b_i, c_j : 1 \leq i \leq p, 1 \leq j \leq p-1\}$ (see Figure 66) (Birman [61], Lickorish [153]).

62. Prove that the geometric coordinates $(\ln l_1, \tau_1, \ln l_2, \tau_2, \dots, \ln l_{3p-3}, \tau_{3p-3})$ can be introduced in the Teichmüller space $T(p, 0), p > 1$, where l_i is the length of a closed geodesic in the homotopy class of the ith loop (see the choice of these loops in Figure 66), and τ_i is the winding coefficient with respect to this geodesic (Thurston [234]).

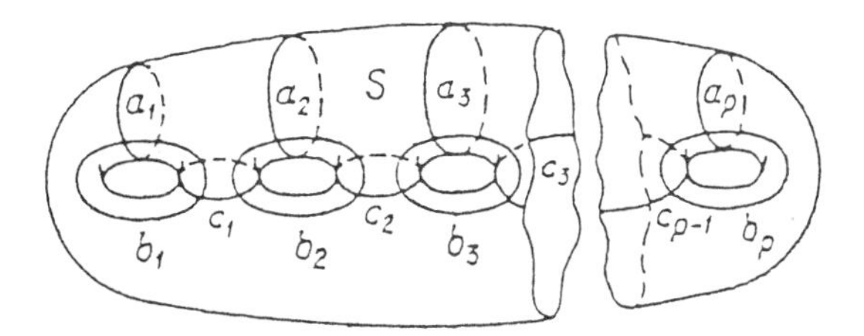

FIGURE 66

63. Let S be a closed Riemann surface of genus $p \geq 2$. Denote by $l(S)$ the length of a shortest closed geodesic on S in the hyperbolic metric, and by $d(S)$ the hyperbolic diameter of S.

Prove the following assertions:

a) $d(S) \leq (p-1)l(S)/\sinh^2(l(S)/2)$, and $l(S)/\sinh^2(l(S)/2) \approx 4/l(S)$ for sufficiently small $l(S)$,

b) there exist constants c_1 and c_2 depending only on the genus p such that $\ln(c_1/l(S)) \leq d(S) \leq 6p\ln(c_2/l(S))$ (Bers [53], Wolpert [248]).

64. Let $c > 0$, and let $K_c \subset R(p,0) = T(p,0)/\operatorname{Mod} T(p,0)$ be the set of closed Riemann surfaces of genus $p \geq 2$ on which each closed geodesic in the hyperbolic metric has length at most c. Prove that K_c is compact in $R(p,0)$ (Bers [53], Mumford [195]).

This gives one way of describing the boundary of $T(p,0)$, since $l(S) \to 0$ as $S \to \partial T(p,0)$.

65. Let $p, p' > 0$ and $n \geq 1$, and let $M_p(p',n)$ be the set of points in the space $R(p,0)$ corresponding to surfaces having a holomorphic mapping onto some surface of genus p' with multiplicity n. Prove that the set $M_p(p',n)$ is an algebraic subset of the space $R(p,0)$ and that its dimension is equal to $(2p-2)-(2n-3)(p'-1)$ (Lange [147]).

66. Prove that for $n = 0$ the Riemann space

$$R(p,0) = T(p,0)/\operatorname{Mod} T(p,0)$$

is simply connected (MacLachlan [161]).

67. Define a *surface of type* (p,n,m) to be a Riemann surface S obtained from a closed surface of genus p by throwing away n points and cutting out m conformal disks ($p \geq 0, n \geq 0, m \geq 0$). The *Teichmüller space* $T(p,n,m)$ of (quasiconformal deformations of) surfaces of type (p,n,m) is defined in a way analogous to that for the case of surfaces without boundary.

A. Prove that there exists a real analytic diffeomorphism of the space $T(p, n, m)$ onto the product of $6p-6+2n+3m$ open intervals (Keen [115]).([20])

It is more precise to speak of surfaces of finite *conformal* type (p, n, m), since punctures and deleted disks are not distinguished topologically.

B. *Nielsen extensions of a Riemann surface with boundary.* Let S be a Riemann surface of finite conformal type (p, n, m), with $m > 0$ and

$$6p - 6 + 2n + 3m > 0.$$

Let us represent S in the form U/Γ, where U is the upper half-plane $\{z: \operatorname{Im} z > 0\}$ and Γ is a torsion-free finitely generated Fuchsian group of the second kind acting discontinuously on the set $\Omega = U \cup L \cup \mathcal{I}$, where L is the lower half-plane and $\mathcal{I}$ is a countable union of open intervals I on the axis $\overline{\mathbf{R}} = \mathbf{R} \cup \{\infty\}$. Then $S^{\mathrm{d}} = \Omega/\Gamma$ is (to within conformal equivalence) the double of the surface S, and complex conjugation $z \mapsto \bar{z}$ induces an anticonformal involution j on S^{d}. The surface S itself is canonically imbedded in S^{d}.

The stabilizer $\Gamma(I)$ in Γ of each of the indicated intervals I is generated by a hyperbolic element with axis $a(I)$ that is a non-Euclidean line in U joining the endpoints of I, and $I/\Gamma(I)$ is homeomorphic to a circle and corresponds to one of the m (ideal) boundary curves of the surface S.

We consider the *Nielsen region* (or *hull*) $N(\Gamma)$ of the group Γ in U, which is obtained from U by removing the regions bounded by I and $a(I)$ for all the indicated intervals I. This is a simply connected region that is convex in the non-Euclidean sense and invariant with respect to Γ. The Riemann surface $S_0 = N(\Gamma)/\Gamma$ is called the *Nielsen kernel* of the surface S (and is canonically imbedded in S). It is a deformation retract of S.

The restriction to S of the hyperbolic metric of the double S^{d} is called the *intrinsic metric* of S; the boundary curves of S are geodesics for this metric.

The following assertions are valid:

(1) *The surface S itself is the Nielsen kernel of some uniquely determined Riemann surface S_1.*

The surface S_1 is called the *Nielsen extension* of S and is obtained geometrically by gluing infinite horns to S along the boundary curves. Further, it is possible to construct an extension S_2 of S_1 and to continue this process indefinitely (each time regarding S_k as imbedded canonically in S_{k+1}). The resulting Riemann surface $S_\infty = S_1 \cup S_2 \cup \cdots$ is called the *infinite Nielsen extension* of S; it does not have boundary curves.

(2) *Every quasiconformal homeomorphism f of S can be extended without increasing the dilation to a quasiconformal homeomorphism of the Nielsen extensions of the surfaces S and $f(S)$ and then to their infinite extensions* (Bers [264]).

([20]) *Translation editor's note*: The analogous definition of the Teimüller space leads to an infinite dimensional space.

C. *The mapping class group of a surface and the Teichmüller modular group.* Let S be a Riemann surface of type (p,n,m) with hyperbolic double S^{d}. We consider the group $\mathrm{Diff}_+(S)$ of orientation-preserving diffeomorphisms of the surface S onto itself (with isolated singularities in general) that extend to quasiconformal diffeomorphisms of the whole double S^{d} (so that punctures pass into punctures). This group is completely determined by the differentiable structure of S and can be written $\mathrm{Diff}_+(p,n,m)$ instead of $\mathrm{Diff}_+(S)$. Let Diff_0 be the normal subgroup of Diff_+ formed by the diffeomorphisms f homotopic to the identity on S. The elements $f \in \mathrm{Diff}_+$ induce in a natural way corresponding topological automorphisms f^* of the Teichmüller space $T(p,n,m)$. Furthermore, f leaves fixed the point $[S] \in T(p,n,m)$ (the equivalence class of marked Riemann surfaces of the type (p,n,m)) if and only if $f \in \mathrm{Diff}_+$ is homotopic to a conformal automorphism of each of the surfaces of the class $[S]$. Let $I = I(p,n,m)$ be the subgroup of Diff_+ that acts trivially on $T(p,n,m)$. The group

$$\mathrm{Diff}_+/\mathrm{Diff}_0 = M_{p,n,m}$$

is called the *class group of* (differentiable) *mappings of the surface* S; the Teichmüller modular group is

$$\mathrm{Mod}(p,n,m) = \mathrm{Diff}_+/I.$$

If we exclude fairly rare cases, then (cf. Problem 1)

$$M_{p,n,m} = \mathrm{Mod}(p,n,m)$$

These exceptional cases are obtained from the description of the elements of $M_{p,n,m}$ of finite order.

Prove the next result, which strengthens and supplements the assertion of Problem 66:

For surfaces of finite type $(p,n) = (p,n,0)$ *the fundamental group of the moduli space* $R(p,n) = T(p,n)/\mathrm{Mod}(p,n)$ *is a cyclic group of order* 5 *when* $p = 2$ *and* $n \equiv 4 \pmod 5$, *and is trivial in all other cases* (Patterson [297]).

D. *The Nielsen-Thurston-Bers classification of diffeomorphisms of a surface of finite type.* The classification of diffeomorphisms of an orientable surface of finite type goes back to work of Nielsen [294] at the end of the 1920's. A complete classification was given by Thurston [302] on the basis of his theory of measured foliations on a surface and his compactification of Teichmüller space. Another approach was proposed by Bers [265] and based on the solution of a certain extremal problem on minimizing the deviation from a conformal structure. Our presentation follows Abikoff's book [251].

Let S be an orientable differentiable surface of type (p,n,m), with

$$6p - 6 + 2n + 3m > 0. \qquad (*)$$

Given a diffeomorphism $\sigma\colon S \to S'$ of S onto a Riemann surface S', it is possible to equip S itself with a conformal structure $S_\sigma = (S,\sigma)$ by lifting it from S' by means of σ. If f is a diffeomorphism of S onto itself, then the quantity $K(\sigma\circ f\circ\sigma^{-1})$ is a measure of the deviation of the mapping $\sigma\circ f\circ\sigma^{-1}$ from a conformal mapping.

By Teichmüller's theorem, the homotopy class $[f]$ of a diffeomorphism f contains precisely one extremal mapping f_σ such that

$$K(\sigma\circ f_\sigma\circ\sigma^{-1}) = \inf K(\sigma\circ f'\circ\sigma^{-1})$$

(over all $f'\in[f]$). This mapping is either conformal or a Teichmüller mapping, i.e., has Beltrami coefficient of the form $k\overline{\varphi(z)}/|\varphi(z)|$, where $0<k<1$, and $\varphi(z)\,dz^2\not\equiv 0$ is an integrable holomorphic quadratic differential on S'.

Following Bers [265], we consider the problem of minimizing the deviation $K(\sigma\circ f\circ\sigma^{-1})$ over all possible conformal structures σ, and define

$$\overline{K}(f) = \inf_\sigma K(\sigma\circ f_\sigma\circ\sigma^{-1}).$$

A conformal structure σ on which $\overline{K}(f)$ is attained is said to be *f-minimal.*

This problem can be reformulated in terms of the Teichmüller metric τ_T of the space $T(p,n,m)$ and the element g_f of the modular group $\mathrm{Mod}(p,n,m)$ induced by the diffeomorphism f, and then we arrive at the result

$$\inf_{S'\in T(S_\sigma)}(S',g_f(S')) = \overline{K}(g_f)$$

(S and S_σ are regarded as *marked* surfaces).

An element $g\in\mathrm{Mod}(p,n,m)$ is said to be *elliptic* if it has a fixed point in $T(p,n,m)=T(S_\sigma)$, *parabolic* if it does not have fixed points and $\overline{K}(g)=0$, *hyperbolic* if $\overline{K}(g)>0$ and there exists a point $S'\in T(S_\sigma)$ with $\tau_T(S',g(S'))=\overline{K}(g)$, and *pseudohyperbolic* if $\overline{K}(g)>0$ and $\tau_T(S',g(S'))>\overline{K}(g)$ for all $S'\in T(S_\sigma)$. This classification is a generalization of the classification of the elements of an elliptic modular group.

We now indicate some properties of diffeomorphisms of a surface of finite type. We first recall that, by a known result of Mangler [287], for such surfaces two homeomorphisms are homotopic if and only if they are isotopic (i.e., homotopic by a homotopy $F(t,x)$ for which all the intermediate mappings are also homeomorphisms).

a) *Periodic mappings and elliptic transformations.* An orientation-preserving diffeomorphism $f\colon S\to S$ is said to be *periodic* if $f^N=\mathrm{id}$ for some $N>0$.

(I) *A diffeomorphism $f\in\mathrm{Diff}_+(S)$ is isotopic to a periodic mapping of S if and only if S has a conformal structure σ and an automorphism f' isotopic to f such that the mapping $\sigma\circ f'\circ\sigma^{-1}$ is conformal.*

(II) *A diffeomorphism f is isotopic to a periodic automorphism of S permuting the punctures on S if and only if the corresponding element $g_f\in\mathrm{Mod}(p,n,m)$ is elliptic.*

(III) *An element* $g \in \operatorname{Mod}(p,n,m)$ *is elliptic if and only if it has finite order.*

(IV) *Let* $S_\sigma \in T(p,n,m)$, $m \neq 0$, *and let* $f \in \operatorname{Diff}_+(S_\sigma)$. *If* f *is not isotopic to a periodic automorphism of* S_σ, *then* $\overline{K}(f) < K(f_\sigma)$.

b) *Irreducible diffeomorphisms and hyperbolic transformations.* Let $\{\alpha_1, \dots, \alpha_r\}$ be a collection of disjoint simple loops on a surface S, where each loop α_j cannot be deformed continuously into a point, nor into a boundary component of S, nor into another loop α_i with $i \neq j$. We say that $\{\alpha_1, \dots, \alpha_r\}$ is a *regular separation* of S if each component of $S \setminus (\bigcup \alpha_j)$ also has nonexceptional type, i.e., satisfies the condition $(*)$.

A diffeomorphism $f \in \operatorname{Diff}_+(S)$ is said to be *reduced* by the collection $\{\alpha_1, \dots, \alpha_r\}$ if $\{\alpha_1, \dots, \alpha_r\}$ is a regular separation of S and $f(\alpha_1 \cup \dots \cup \alpha_r) = \alpha_1 \cup \dots \cup \alpha_r$; automorphisms of S isotopic to such f are said to be *reducible*. An f is said to be *irreducible* if it is not isotopic to any reducible diffeomorphism. For example, a Dehn twist around a homotopically nontrivial loop α is a reducible diffeomorphism.

The following assertions hold for surfaces of type $(p,n,0) = (p,n)$:

(V) *If* $f \in \operatorname{Diff}_+(S)$ *is irreducible, then the element* $g_f \in \operatorname{Mod}(p,n,0)$ *corresponding to it is hyperbolic or elliptic.*

(VI) *If* g_f *is hyperbolic, then there is an infinite set of distinct conformal structures on which* $\overline{K}(f)$ *is attained.*

The proofs of these and other important properties of diffeomorphisms of a surface are based on an idea in differential geometry relating to straight line spaces (in the sense of Busemann). The role of the straight lines in $T(p,n,0) = T(S_0)$ *is played by the Teichmüller straight lines* $t \mapsto S_t$, *where* $t = (1+\xi)/(1-\xi)$, $-1 < \xi < 1$, *and* S_t *is obtained from* S_σ *by a quasiconformal Teichmüller deformation* $w^{|\xi|\bar{\omega}/\omega}$, $\omega \not\equiv 0$ *being a holomorphic quadratic differential on* S_0.

(VII) *Let* S_σ *be a Riemann surface of type* $(p,n,0)$ *with* $2p-2+n > 0$. *Suppose that* $f \in \operatorname{Diff}_+(S_\sigma)$ *and that the element* g_f *of* $M(p,n,0)$ *corresponding to it has infinite order. Then* g_f *is hyperbolic if and only if there exists a straight line* $L \subset T(p,n,0)$ *passing through* S_σ *which is carried by* g_f *into itself:* $g_f(L) = L$.

c) *Pseudo-Anosov diffeomorphisms.* Hyperbolic transformations of the group $\operatorname{Mod}(p,n,0)$ are induced by diffeomorphisms of the surface with definite dynamical properties.

Each nonzero integrable holomorphic quadratic differential $\varphi\, dz^2$ on a Riemann surface S_σ of type $(p,n,0)$ determines on S two transversal measured foliations: the *horizontal* and *vertical* foliations consisting, respectively, of the lines on which $\varphi > 0$ and $\varphi < 0$ (these foliations have finitely many singularities at the zeros of φ).

Suppose now that S is a surface of finite topological type $(p,n,0)$. A diffeomorphism $f\colon S \to S$ is called an *Anosov diffeomorphism* if for some conformal structure $S_\sigma = (S,\sigma)$ on S there exist a pair $(\mathcal{F}_1, \mathcal{F}_2)$ of transversal measured foliations on S and a number K such that the following conditions hold:

(i) The image of a leaf of $\mathcal{F}_j$ is a leaf of $\mathcal{F}_j$, $j = 1, 2$.

(ii) The distances between leaves ψ_1^j and ψ_2^j of $\mathcal{F}_j$ satisfy the equalities

$$\delta(f(\psi_1^1), f(\psi_2^1)) = K^{1/2}\delta(\psi_1^1, \psi_2^1),$$
$$\delta(f(\psi_1^2), f(\psi_2^2)) = K^{-1/2}\delta(\psi_1^2, \psi_2^2).$$

A differential f with isolated singularities (the set of these singularities is denoted by b_f) is called a *pseudo-Anosov diffeomorphism* if

i) f is an Anosov diffeomorphism on $S \backslash b_f$, and

ii) the foliations $\mathcal{F}_j$ have identical singularities at the points $z \in b_f$, and these singularities are homeomorphic to the singularities of the horizontal foliation connected with the quadratic differential.

In the framework of the Teichmüller theory of extremal quasiconformal mappings this definition has the following meaning. As mentioned above, each extremal (in the Teichmüller sense) quasiconformal mapping F of a Riemann surface X of finite type $(p, n, 0)$ has Beltrami coefficient of the form $k\overline{\varphi(z)}/|\varphi(z)|$, where $0 < k < 1$, and $\varphi(z)\,dz^2 \not\equiv 0$ is an integrable holomorphic quadratic differential on X which is uniquely determined if, for example, it is required that $\int\int_X |\varphi|\,dx\,dy = k$ $(z = x + iy)$. Then there is a unique holomorphic quadratic differential $\psi(w)\,dw^2$ on the surface $F(X)$ such that, in the local parameters

$$\varsigma = \xi + i\eta = \int \varphi^{1/2}\,dz, \qquad \varsigma' = \xi' + i\eta' = \int \psi^{1/2}\,dw,$$

the mapping F has the form

$$\varsigma' = K^{1/2}\xi + K^{-1/2}\eta \qquad (K = (1 + k)/(1 - k))$$

at points where φ is nonzero, and

$$\iint\limits_X |\varphi|\,dx\,dy = \iint\limits_{F(X)} |\psi|\,du\,dv \qquad (w = u + iv).$$

The differentials φ and ψ are called, respectively, the *initial* and *terminal differentials* for the diffeomorphism F.

A pseudo-Anosov diffeomorphism of a differentiable surface S is a mapping that can be written as $\sigma \circ F \circ \sigma^{-1}$, where σ is a mapping of S onto a Riemann surface X without boundary curves, and $F\colon X \to X$ is a Teichmüller mapping for which the initial and terminal quadratic differentials coincide.

THURSTON'S THEOREM. *If S is a surface of type $(p, n, 0)$, and the differential $f \in \mathrm{Diff}_+(S)$ is not isotopic to a periodic automorphism of S, then either* (i) *f is isotopic to some reducible automorphism or* (ii) *f is isotopic to a pseudo-Anosov diffeomorphism and the corresponding element*

$g_f \in \mathrm{Mod}(p, n, 0)$ *is hyperbolic. Furthermore, cases* (i) *and* (ii) *cannot hold simultaneously.*

The reader interested in the topological and dynamical aspects of Thurston's theory can turn to the book [269].

E. Let g be a hyperbolic modular transformation of a finite-dimensional Teichmüller space $T(\Gamma) = T(p, n)$. Prove that if $\lim_{j\to\infty} g^{m_j} = \hat{g}$ exists for some infinite sequence of iterates g^{m_j}, then $\hat{g}(T(\Gamma)) = \varphi$, where φ is a boundary point of the complex boundary $\partial T(\Gamma)$ (in $B(L, \Gamma)$), and the group Γ_φ corresponding to this point φ is degenerate (Bers [266]).

F. *Fenchel-Nielsen coordinates in Teichmüller space.* Along with complex coordinates determining the analytic structure on the Teichmüller space $T(p, n)$, it is sometimes more convenient to parametrize the points of this space by other coordinates (cf. Problem 62). Coordinates on marked Riemann surfaces called *Fenchel-Nielsen coordinates* have proved to be useful in studying the compactification of Teichmüller spaces and in questions connected with the classification of diffeomorphisms of a surface of finite type. They are the lengths of ideal boundary curves of certain three-connected regions and the sizes of certain conformal gluing angles.

First of all, each surface S of nonexceptional type (p, n, m) can be decomposed into so-called *pants* that are surfaces S_0 with boundary whose interior is homeomorphic to the three-connected region

$$\{z \in \mathbf{C}\colon |z| < 4,\ |z-2| > 1,\ |z+2| > 1\}$$

(see Figure 67). Pants for which some of the boundary components C degenerate into a puncture are said to be *contracted around* C. The boundary components of pants S_0 are conveniently called the *belt* W, the *left cut* L, and the *right cut* R. The fundamental group $\pi_1(S_0)$ is a free group with two generators.

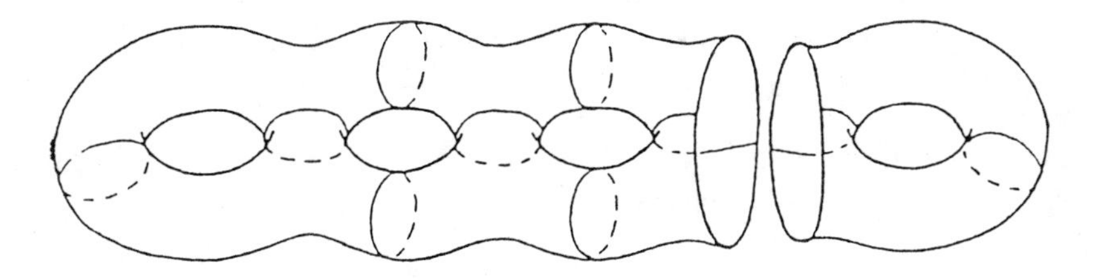

FIGURE 67

It can be proved that *the conformal structure of pants S_0 is determined to within conformal or anticonformal equivalence by the triple of numbers $l\{W\}, l(L)$, and $l(R)$, where $l\{\alpha\}$ is the hyperbolic length of a geodesic in the homotopy class $[\alpha]$ of a closed curve $\alpha \in S_0$.*

The following assertion is obtained from this by using the Nielsen kernel of the surface S_0.

Let S_1 be pants. For a closed curve α on S_1 let $l_I(\alpha)$ denote the length of α in the intrinsic metric on S_1 (see Problem 67B) and let

$$l_I\{\alpha\} = \inf\{l_I(\alpha')\colon \alpha' \in [\alpha]\}.$$

Then the surface S_1 is uniquely determined by the quantities $l_I\{W\}$, $l_I\{L\}$, and $L_I\{R\}$ to within conformal or anticonformal equivalence.

We now show how to use a decomposition of a Riemann surface into pants for parametrizing a Teichmüller space.

Let S be a Riemann surface of type (p, n, m) with $6p-6+2n+3m > 0$. A decomposition $\mathcal{P}$ of S into pants is called a *maximal* (geodesic) *decomposition* if each component of $\partial_S S_i$ is a simple geodesic loop.

Denote the boundary components of S_i by C_{ik}, with the convention that $C_{i,4} = C_{i,1}$. For each pair (i, k) there is a unique arc α that is a geodesic with respect to the intrinsic metric and such that:

(a) one end of α lies on $C_{i,j}$, and the other on $C_{i,j+1}$;

(b) if $C_{i,j}$ and $C_{i,j+1}$ are boundary curves, then $l_I(\alpha)$ is minimal among the lengths of all the arcs satisfying (a); and

(c) if $C_{i,j}$ (respectively, $C_{i,j+1}$) is a puncture, and $C_{i,j+1}$ (respectively, $C_{i,j}$) is a boundary curve, then α is orthogonal to $C_{i,j+1}$ (respectively, $C_{i,j}$).

The endpoint of α on the curve $C_{i,j}$ is denoted by $\varsigma_{i,j}$ and regarded as its base point on $C_{i,j}$. When $C_{i,j}$ is identified with $C_{i',j'}$, we choose the positive direction on $C_{i,j}$ compatible with this identification. If $\mathcal{P}$ decomposes S in such a way that $C_{i,j}$ is thereby identified with $C_{i',j'}$, then we identify these curves so that the point $\varsigma_{i,j}$ coincides with $\varsigma_{i',j'}$. In general this gives a new Riemann surface S_0. On S_0 we choose some marking (see §5 in Chapter II).

Note that if C is a rectifiable curve on S_i with intrinsic length $l_I(C)$, then the length of C in the intrinsic metric of S_0 is also equal to $l_I(C)$. In particular, if S_0 is a surface of type $(p, n, 0)$, then $l_I(C)$ is the hyperbolic length of C. Note also that the requirement that the points $\varsigma_{i,j}$ and $\varsigma_{i',j'}$ coincide under the identification of $C_{i,j}$ and $C_{i',j'}$ can be removed. Then we arrive at another sum S_0' of the surfaces S_i.

It is assumed that $i < i'$, or $j < j'$ if $i = i'$. Let $\delta_{i,j}$ be the directed intrinsic distance from $\varsigma_{i,j}$ to $\varsigma_{i',j'}$, and define

$$\theta_{i,j} = 2\pi\delta_{i,j}/l_I(C_{i,j})$$

(here $\theta_{i,j}$ is uniquely determined modulo 2π).

The intrinsic length of the curves $C_{i,j}$ can be changed; then the conformal structure of the pants S_i also changes. If we do this while coordinating with the subsequent identification of boundary curves, then, as earlier, we can perform a conformal gluing.

We consider the complete collection L of intrinsic lengths of the curves $C_{i,j}$, with the following conventions:

(a) $l_I(C_{i,j})$ precedes $l_I(C_{i',j'})$ if $i < i'$, or if $i = i'$ but $j < j'$.

(b) $l_I(C_{i,j})$ is omitted if $C_{i',j'}$ is identified with $C_{i,j}$ and $i < i'$, or if $j < j'$ when $i = i'$.

(c) $l_I(C_{i,j})$ is omitted if $C_{i,j}$ is a puncture.

(d) The lengths of the dissecting curves precede the lengths of the boundary curves.

Further, let Θ be the collection of angles $\theta_{i,j}$ also subordinate to conditions of the type (a) and (b). Moreover, we omit $\theta_{i,j}$ if $C_{i,j}$ is an ideal boundary component of S_0. The collections L and Θ can be regarded as sets of functions on $T(S_0)$. Further, a maximal decomposition of a surface of nonexceptional type (p, n, m) consists of $2p - 2 + m + n$ elements, so that L consists of $3p - 3 + 2m + n$ elements, and Θ consists of $3p - 3 + m + n$ elements (since the m angles $\theta_{i,j}$ corresponding to the boundary curves of S_0 are omitted).

In place of L it is more convenient to take the collection of numbers inverse to the quantities $l_I(C_{i,j})$; denote it by L^{-1}. Then we get the mapping

$$(L^{-1}, \Theta)\colon T(S_0) \to \mathbf{R}_+^{3p-3+2m+n} \times (S^1)^{3p-3+m+n}.$$

THEOREM. *The mapping (L, Θ) is a real analytic covering.*

A proof can be found in Abikoff's book [251]. It is a translation of van Kampen's theorem to the geometric language of a universal covering and actually follows the Maskit combination theorem, applied to Fuchsian groups.

As a consequence we get that *the mapping*

$$(L^{-1}, \boldsymbol{\Theta})\colon T(S_0) \to \mathbf{R}_+^{3p-3+2m+n} \times \mathbf{R}^{3p-3+m+n},$$

which covers (L^{-1}, Θ) is a real analytic diffeomorphism.

The parameters $(L^{-1}, \boldsymbol{\Theta})$ are called *Fenchel-Nielsen coordinates* of a surface $S \in T(S_n)$. The coordinates $(L^{-1}, \boldsymbol{\Theta})$ and (L^{-1}, Θ) depend on the initial partition of S_0 into pants.

These coordinates can be connected with the Dehn-Lickorish twists in Problem 62. Confining ourselves here to the case of closed surfaces of genus $p > 1$, we note that

$$(L^{-1}, \Theta)(T(S_0)) = T(S_0)/\mathcal{T},$$

where $\mathcal{T}$ is a free abelian group generated by the twists with respect to all the closed homotopically nontrivial loops α indicated in Figure 67.

68. Prove that a) all finite-dimensional Teichmüller spaces $T(\Gamma)$ are contractible and, consequently, all the groups $\pi_i(T(\Gamma))$ are 0 for $i \geq 1$; and b) the universal space $T(\mathbf{1})$ is also contractible (Earle and Eells [71]).

69. Let $a(X) = (a_1, \dots, a_m)$ be arbitrary local coordinates (moduli) in a neighborhood of a point $X_0 \in T(p, n)$, $m = 3p-3+n$ $(a(X_0) = 0)$, that define a complex structure there. Prove that

$$a_j(X) = \iint_{X_0} \mu(z)\varphi_j(z)\,dx\,dy + O(\|\mu\|^2), \qquad j = 1, \dots, m, \tag{117}$$

where μ is the Beltrami differential of the quasiconformmal mapping $f^\mu\colon X_0 \to X = X_0^\mu$, and $\varphi_1, \dots, \varphi_m$ is a corresponding basis in the space of integrable holomorphic quadratic differentials on X_0, and that the estimate of the remainder term in (117) is uniform for $\|\mu\| \le k < 1$ (Krushkal′ [130]).

70. (Other global moduli of Riemann surfaces of finite type). Let Γ be a finitely generated torsion-free Fuchsian group of the first kind with limit circle $|z| = 1$, let U^* be the region $\{z \in \overline{\mathbf{C}}\colon |z| > 1\}$, and let $f_\mu(z)$ be quasiconformal automorphisms of $\overline{\mathbf{C}}$ with $\mu \in M(U^*, \Gamma)$, normalized by the conditions $f_\mu(0) = 0$, $f'_\mu(0) = 1$ and $f_\mu(\infty) = \infty$. Assume also that $z = 0$ is not a *Weierstrass point for the group* Γ, i.e., there does not exist a nonzero function $\varphi \in B(U, \Gamma)$ such that $\varphi(0) = \varphi'(0) = \cdots = \varphi^{(m-1)}(0)$, where $m = \dim B(U, \Gamma)$ (if this is not so, then it is possible to pass from Γ to a conjugate group $\alpha\Gamma\alpha^{-1}$ for which $z = 0$ is no longer a Weierstrass point by means of a Möbius automorphism α of U).

Consider the Taylor expansion

$$f_\mu(z) = z + \sum_{k=2}^{\infty} a_k z^k, \quad \{f_\mu, z\} = \sum_{k=0}^{\infty} b_k z^k, \qquad z \in U. \tag{118}$$

Prove that if U/Γ is a surface of type (p, n), i.e., $m = 3p - 3 + n$, then the first $m + 1$ coefficients a_k or the first m coefficients b_k in (118) can be taken as moduli of the points in $T(p, n)$. More precisely:

a) the points $a = (a_2, \dots, a_{m+2})$ fill a bounded region on an $(m+1)$-dimensional analytic surface in $\mathbf{C}^{m+1}$, and the region is biholomorphically equivalent to the space $T(p, n)$; and

b) the points $b = (b_0, \dots, b_{m-1})$ fill a bounded region in $\mathbf{C}^m$ that is biholomorphically equivalent to $T(p, n)$ (Krushkal′ [130], Lyan [157]).

71. Let $T(\Gamma) = T(p, n)$, and let $\tilde{T}(\Gamma)/\Gamma = V(p, n)$, which is a fibering into Riemann surfaces of type (p, n) over the space $T(\Gamma)$ and is also called a *Teichmüller curve*; recall that the projection $\pi_n\colon V(p, n) \to T(p, n)$ is holomorphic (see §5 in Chapter II). Prove the following assertions:

a) $\pi_n\colon V(p, n) \to T(p, n)$ has exactly n global holomorphic sections when $p \ge 3$, and $2n + 6$ holomorphic sections when $p = 2$ $(n \ge 0)$; the sections corresponding to the Weierstrass points of the Riemann surfaces correspond to the case $n = 0$ $(p = 2)$, and the remaining sections (for $n > 0$) correspond to the n deleted points.

b) The curves $\pi_n\colon V(1, n) \to T(1, n)$ for $n \ge 3$ and $\pi_n\colon V(0, n) \to T(0, n)$ for $n \ge 5$ have n holomorphic sections (Hubbard [103], Earle and Kra [72]).

On Hermitian metrics on the space $T(p, n)$. The Teichmüller (Kobayashi) metric in $T(p, n)$ is not Hermitian. To obtain Hermitian metrics one should, as usual, consider positive-definite Hermitian inner products on the tangent or cotangent spaces at points of $T(p,n)$. In particular, the cotangent space at a point $X_0 \in T(p,n)$ consists of integrable holomorphic quadratic differentials $\varphi\, dz^2$ on the Riemann surface X_0 (more precisely, a representative of this equivalence class), and by taking some conformal metric $\lambda(z)|dz|$ on X_0 one obtains a Hermitian inner product

$$(\varphi_1, \varphi_2) = \iint_{X_0} \lambda^{-2}(z)\varphi_1(z)\overline{\varphi_2(z)}\, dx\, dy. \tag{119}$$

Choosing the $\lambda(z)$ in (119) to be the usual hyperbolic metric, we arrive at the so-called *Weil-Petersson metric* $w_T(x_0, x)$. Namely, let X_0 be represented in the form $X_0 = H/\Gamma_0$, and let $\varphi_1, \ldots, \varphi_m$ be a basis in $B(H, \Gamma)$ such that

$$\iint_{H/\Gamma_0} (2y)^2 \varphi_j(z)\overline{\varphi_k(z)}\, dx\, dy = \delta_{jk}.$$

A neighborhood of X_0 in $T(p,n)$ is filled by points $X = X_0^\mu$ with $\mu = (2y)^2 \sum_{j=1}^m \xi_j \overline{\varphi_j(z)}$, and the points $\xi_1, \ldots, \xi_m$ are taken as local coordinates in this neighborhood. Then the differential form corresponding to the metric w_T is $ds^2 = \sum_j |d\xi_j|^2$.

72. Prove that the Weil-Petersson metric is Kählerian, has negative holomorphic curvature, and is not complete, and that there exists a constant $c_{p,n}$, depending only on $T(p,n)$, such that $w_T(X_0, X) \le c_{p,n} d(X_0, X)$ (see Weil [242], Ahlfors [14], Linch [154], and Wolpert [248]).

73. The projection $\pi_p \colon T(p,0) \to R(p,0) = T(p,0)/\operatorname{Mod} T(p,0)$ enables one to define in $R(p,0)$ the metric

$$\tilde{w}_R(\tilde{X}_1, \tilde{X}_2) = \inf w_T(\pi_p^{-1}(\tilde{X}_1), \pi_p^{-1}(\tilde{X}_2)).$$

Prove that the diameter of the space $R(p,0)$ in the metric $\tilde{w}_R$ is finite (Wolpert [248]).

74. Let Γ be an arbitrary Fuchsian group, and let $c_T(\cdot,\cdot)$ be the *Carathéodory metric* in the Teichmüller space $T(\Gamma)$, i.e.,

$$c_T(X_1, X_2) = \sup \rho(f(X_1), f(X_2))$$

over all holomorphic mappings $f \colon T(\Gamma) \to U$, where ρ is the hyperbolic metric in the disk U.

a) Prove that the metric $c_T(\cdot,\cdot)$ is complete for any $T(\Gamma)$, and if $\dim T(\Gamma) < \infty$, then $c_T(\cdot,\cdot)$ is even strongly complete (i.e., any ball $\{X \in T(\Gamma) \colon c_T(X_0, X) \le r < \infty\}$ is compact in $T(\Gamma)$) (Krushkal′, Earle; see [133] and [137]).

It is known that this implies, in particular, that every holomorphic mapping $f \colon U \backslash \{0\} \to T(\Gamma)$ extends to a holomorphic mapping of the whole disk U.

b) Prove that the Carathéodory metric of a space of dimension greater than 2 does not coincide with its Teichmüller-Kobayashi metric (Krushkal′ [281], [282]).

75. Coverings $\pi\colon T \to S$ and $\pi'\colon T' \to S$ are said to be *equivalent* if there exists a homeomorphism $h\colon T \to T'$ such that $\pi = \pi' \circ h$. Prove that there are precisely 100 nonequivalent three-sheeted coverings over a pretzel, of which 40 are regular and the rest are not (Mednykh [193]).

76. Prove that the number of subgroups of index n in the fundamental group of a closed Riemann surface of genus p is

$$M_p(n) = n \sum_{s=1}^{n} \frac{(-1)^{s+1}}{s} \sum_{\substack{i_1+i_2+\cdots+i_s=n \\ i_1, i_2, \ldots, i_s \geq 1}} \beta_{i_1}\beta_{i_2}\cdots\beta_{i_s}, \tag{120}$$

where $\beta_k = \sum_{\lambda \in D_k} (k!/f^{(\lambda)})^{2p-2}$, D_k is the set of all irreducible representations of the symmetric group S_k, and $f^{(\lambda)}$ is the degree of the representation λ (Mednykh [193]).

77. Prove that the number of nonequivalent n-sheeted coverings over a closed Riemann surface of genus p is

$$N_p(n) = \frac{1}{n} \sum_{m/n} M_p(m) \sum_{d|n/m} \mu\left(\frac{n}{md}\right) d^{(2p-2)m+2}, \tag{121}$$

where $\mu(n)$ is the number-theoretic Möbius function, and $M_p(n)$ is the same as in Problem 76.

Deduce from this that $M_1(n) = \sum_{d|n}^{n} d$ (the sum of the divisors of a number n), and then in (120) and (121) we have that $M_p(n) \sim 2(n!)^{2p-2}, n > 1$, as $p \to \infty$, and $\sum_{s=1}^{n} M_1(s) \sim \pi^2 n^2/12$ as $n \to \infty$ (Mednykh).

78. (On surfaces with nodes). A *Riemann surface with nodes* is defined to be a connected one-dimensional complex analytic space S, each point P of which has a neighborhood that is (holomorphically) isomorphic either to the disk or to the set $\{z = (z_1, z_2) \in \mathbf{C}^2\colon z_1 z_2 = 0, |z_1| < 1, |z_2| < 1\}$; in the second case P is called a *node*. The simplest example of such a surface is obtained from a thrice-punctured sphere by identifying two of the points deleted (which give the node).

A hyperbolic metric is defined on such surfaces to be a hyperbolic metric on the components of the complement $S\backslash\mathcal{N}(S)$ of the set $\mathcal{N}(S)$ of nodes; each of these components is an ordinary Riemann surface. In a corresponding way one defines topological and quasiconformal deformations of surfaces with nodes, along with the equivalence classes of these deformations, which form the *completed Teichmüller space* $\hat{T}(S)$.

Note that in the preceding definitions it is not required that $\mathcal{N}(S)$ be nonempty; hence ordinary Riemann surfaces can be regarded as a particular case of surfaces with nodes. Therefore, the space $\hat{T}(S)$ can also be obtained directly as the completion of an ordinary Teichmüller space. Namely, let S be a closed Riemann surface of genus $p > 1$ (or even of finite type) represente by a Fuchsian group Γ of the first kind; then $\hat{T}(S)$ is obtained by adjoining to $T(\Gamma) = T(S)$ the part of the boundary $\partial T(\Gamma)$ corresponding to the nondegenerate b-groups.

The space $\hat{R}(S) = \hat{T}(\Gamma)/\operatorname{Mod} T(\Gamma)$, called the *completed Riemann space*, serves as a compactification of the Riemann space $R(S) = T(\Gamma)/\operatorname{Mod} T(\Gamma)$. This is a compact Hausdorff space which is a normal complex space and even a projective algebraic variety. It coincides with the space M_p of moduli of a stable algebraic curve of genus p (see, for example, Abikoff [7], Bers [56] or Mumford [196]).

The space of quasiconformal deformations of an arbitrary Kleinian group on the plane. Let $G \subset \mathcal{M}_2$ be a Kleinian group of finite type, and let $\Delta \subset \Omega(G)$ be some invariant union of components of it:

$$\Delta/G = S_1 \cup S_2 \cup \cdots \cup S_r \qquad (r < \infty). \tag{122}$$

Denote by $M(\Delta, G)$ the Banach space of Beltrami G-differentials equal to zero in $\overline{\mathbf{C}} \backslash \Delta$, with the ordinary norm $\|\mu\|_M = \|\mu\|_\infty$, and let $M_1(\Delta, G)$ be its unit ball and $Q(\Delta, G)$ the set $\{f^\mu\colon \mu \in M_1(\Delta, G)\}$ (regarded as a group under superposition). We introduce in $M_1(\Delta, G)$ two subsets: the set $M_t(\Delta, G)$ consisting of *trivial* differentials μ such that

$$f^\mu(z)|_{\Delta(G)} = I, \tag{123}$$

and the set $\hat{M}_t(\Delta, G)$ of differentials such that, in addition to (123), the quasiconformal homeomorphisms induced on the surfaces S_j in (122) by the automorphisms f^μ are homotopic to the identity on each of them (cf. §5 in Chapter II). The (normal) subgroups of $Q(\Delta, G)$ corresponding to such μ are denoted by $Q_0(\Delta, G)$ and $\hat{Q}_0(\Delta, G)$.

Corresponding to this, the following two spaces of quasiconformal deformations of the group G are defined:

$$T(\Delta, G) = Q(\Delta, G)/Q_0(\Delta, G), \qquad \hat{T}(\Delta, G) = Q(\Delta, G)/\hat{Q}_0(\Delta, G).$$

In particular, $\hat{T}(U, \Gamma) = T(U, \Gamma) = T(\Gamma)$ for a Fuchsian group Γ of the first kind of the disk U. Further, let

$$\mathcal{G}(\Delta, G) = M_t(\Delta, G)/\hat{M}_t(\Delta, G). \tag{124}$$

We take the so-called Fuchsian equivalent of G. Namely, let $\Omega_1, \ldots, \Omega_r$ be a complete set of nonconjugate components of Δ, i.e., components such that $\Omega_j/G_{\Omega_j} = S_j$, $j = 1, \ldots, r$, in (122). Consider the Fuchsian group Γ_j determined by the universal holomorphic covering $h_j\colon U \to \Omega_j$, $j = 1, \ldots, r$.

79. Prove the following assertions:

a) There is a holomorphic isomorphism

$$\hat{T}(\Delta, G) = T(\Gamma_1) \times \cdots \times T(\Gamma_r) \tag{125}$$

(and, in particular, $\hat{T}(\Delta, G)$ is a complex analytic manifold).

b) $T(\Delta, G)$ has the decomposition

$$T(\Delta, G) = T(\Omega_1, G_{\Omega_1}) \times \cdots \times (\Omega_r, G_{\Omega_r}). \tag{126}$$

c) The group $\mathcal{G}(\Delta, G)$ (defined by (124)) acts freely on $\hat{T}(\Delta, G)$.

d) $\hat{T}(\Delta, G)$ is the universal covering space for $T(\Delta, G)$, with

$$T(\Delta, G) = \hat{T}(\Delta, G)/\mathcal{G}(\Delta, G) \tag{127}$$

and, by c), the space $T(\Delta, G)$ is also a complex analytic manifold.

e) If all the components of Δ are simply connected, then $\mathcal{G}(\Delta, G) = 1$, and $T(\Delta, G) = \hat{T}(\Delta, G)$.

The decomposition (126) remains in force also for arbitrary Kleinian groups not necessarily of finite type, i.e., also when there are infinitely many nonconjugate components $\Omega_j \subset \Delta$; the relations (125) and (127) are based essentially on the Maskit extension theorem (see Problem 39) and, hence, on the finiteness of r (Bers [55], Kra [120], Maskit [182]).

By analogy with the Riemann space $R(\Gamma)$ we also introduce the space $\hat{R}(\Delta, G)$ of conjugacy classes in $\mathcal{M}_2$ of Kleinian groups quasiconformally equivalent to G. Moreover, $\hat{R}(\Delta, G) = \hat{T}(\Delta, G)/\operatorname{Mod}\hat{T}(\Delta, G)$, where $\operatorname{Mod}\hat{T}(\Delta, G)$ is a certain (discontinuous) group of holomorphic automorphisms of $\hat{T}(\Delta, G)$; hence $\hat{R}(\Delta, G)$ is a normal complex space.

In the set of Schottky groups of a particular genus $p > 1$ we introduce a conformal equivalence and define the topology of convergence on generators. The space obtained is called *the Schottky space of genus p*. This space admits a complex structure and then becomes a domain of holomorphy; the Teichmüller space $T(p, 0)$ serves as its universal covering (concerning these results and generalizations of them see Hejhal [98] and Chushev [68]).

80. Prove that all the boundary groups of a classical Schottky space (see Problem 25) are geometrically finite (Jørgensen, Marden and Maskit [113]).

81. Prove that a classical Schottky space of genus $p > 1$ is connected (Zarrow [249]).

82. Let $\Gamma \subset \mathcal{M}_2$ be a Fuchsian group, and let $T(\Gamma)$ be its Teichmüller space, which is biholomorphically immersed in $B(L, \Gamma)$. Define the set $S(\Gamma) = S \cap B(L, \Gamma)$, where S is the set of Schwarzian derivatives of functions univalent in L. Prove the following facts:

a) If $f\colon U \to B(L, \Gamma)$ is a mapping holomorphic in U and continuous in $\overline{U}$ and if $f(\partial U) \subset S(\Gamma)$, then $f(U) \subset S(\Gamma)$, and if $f(U) \cap T(\mathbf{1}) \neq \varnothing$, then $f(U) \subset T(\mathbf{1})$, while if $f(U) \cap T(\Gamma) \neq \varnothing$, then $f(U) \subset T(\Gamma)$.

b) The component of zero in the subset $\operatorname{int} S(\Gamma)$ of $B(L, \Gamma)$ coincides with $T(\Gamma)$ (Zhuravlev [250]).

83. Prove that two closed hyperbolic manifolds M_1^n and M_2^n ($n > 2$) with isomorphic fundamental groups are isometric (Margulis [172]).

84. Prove that two complete hyperbolic manifolds M_1^n and M_2^n (of finite volume), $n > 2$, are homeomorphic if and only if they are homotopy equivalent (Thurston [234]).

85. Prove that a hyperbolic dodecahedron space M has finite first homology group $H_1(M, Z)$, but it has a finite-sheeted covering with infinite $H_1(M, Z)$ (Vinberg [236]).

86. Prove that if l is an m-component link on the sphere S^3 ($m \geq 1$), and $\varphi\colon \pi_1(S^3 \backslash l) \to \mathcal{M}_2$ is an anti-isomorphism of the nonabelian fundamental group $\pi_1(S^3 \backslash l)$ onto a discrete group $G \subset \mathcal{M}_2$ that is not discontinuous, i.e., an isomorphism under which corresponding products have reversed order of factors and generators are carried into parabolic mappings, then the complement $S^3 \backslash l$ of the link l admits a complete hyperbolic structure (and has finite volume) and is homeomorphic to the space $\mathbf{R}^3_+/\hat{G}$ of orbits of the extension $\hat{G}$ of G to $\mathbf{R}^3_+$ (Riley [211]).

Let M be a complete hyperbolic manifold (possibly with infinite volume). For $\varepsilon > 0$ define the decomposition $M = M_{(0,\varepsilon]} \cup M_{[\varepsilon,\infty)}$, where $M_{(0,\varepsilon]}$ consists of the points of M through which nontrivial loops of length $\leq \varepsilon$ pass, and $M_{[\varepsilon,\infty)}$ consists of the points $x \in M$ such that any nontrivial loop passing through x has length $\geq \varepsilon$.

87. Use the lemma of Margulis (see Problem 10) to prove that there exists an $\varepsilon > 0$ such that any component $M_{(0,\varepsilon]}$ of a complete orientable hyperbolic 3-manifold M is either a) $B/\mathbf{Z}$ or $B/\mathbf{Z} \oplus \mathbf{Z}$, where B is a horoball, or b) $B_r(\gamma)/\mathbf{Z}$, where $B_r(\gamma)$ is the r-neighborhood of a geodesic γ ($r \geq 0$) (Thurston [234]).

88. Prove that for any constant $c > 0$ there are only finitely many topological types $M_{[\varepsilon,\infty)}$, where M is a complete hyperbolic 3-manifold with $\operatorname{vol} M \leq c$.

In other words, there exists a link $L_c \subset S^3$ such that each complete hyperbolic manifold M with $\operatorname{vol} M \leq c$ can be obtained by Dehn surgery on the link L_c (the limiting case when the result is noncompact is also included (cf. Example 70)) (Jørgensen; see [234]).

89. Let M^n be a complete hyperbolic manifold with $\operatorname{vol} M^n < \infty$, $n \geq 3$. Prove that the group $\operatorname{Out} \pi_1(M^n)$ of outer automorphisms of $\pi_1(M^n)$ is finite and isomorphic to the group of isometries of the manifold M^n (Thurston [234]).

We remark that for $n = 2$ the group $\operatorname{Out} \pi_1(M^2)$ is always infinite.

90. Prove that there exists a constant $r = r(n) > 0$ such that $\operatorname{vol}(M^n) \geq r$ for any complete hyperbolic n-manifold M^n (Apanasov [30]).

91. Prove that the set of volumes of complete hyperbolic 3-manifolds is well ordered (Gromov; see [234]).

92. Prove that any complete three-dimensional hyperbolic manifold M with doubly generated fundamental group $\pi_1(M) = \{a, b\}$ admits an involution carrying a into a^{-1} and b into b^{-1}; if a and b are conjugate, then there exist involutions acting on M like $\mathbf{Z}_2 \oplus \mathbf{Z}_2$ and carrying a into b (if a and b do not have a common fixed point at ∞) (Thurston [234]).

93. Prove that if $a, b \in \pi_1(M)$ and $[a, b] = \mathbf{1}$, where M is a complete hyperbolic 3-manifold, then either 1) a and b are elements of an infinite cyclic subgroup $\{x\} \subset \pi_1(M), x^m = a, x^k = b$, or 2) M has an end E homeomorphic to $T^2 \times [0, \infty)$; moreover, the group $\{a, b\}$ is conjugate in $\pi_1(M)$ to a subgroup of finite index in $\pi_1(E)$ (Thurston [234]).

94. Let M be a three-dimensional compact orientable irreducible manifold such that each component of the boundary is a torus. Prove that $\operatorname{int} M$ has a complete hyperbolic structure if and only if each free abelian subgroup of $\pi_1(M)$ with rank 2 is *peripheral*, i.e., conjugate in $\pi_1(M)$ to the subgroup H that is the image of $\pi_1(S)$ in $\pi_1(M)$ under the homomorphism induced by the imbedding $S \subset M$ for some component $S \subset \partial M$ (Thurston [234]).

95. Let M be a compact three-dimensional manifold with nonempty boundary such that $\operatorname{int} M$ admits a geometrically finite metric with constant negative curvature, i.e., $\operatorname{int} M$ is homeomorphic to $\mathbf{R}^3_+/G$, where G is a geometrically finite torsion-free Kleinian group. Assume that for every component $S \subset \partial M$ the homomorphism $\pi_1(S) \to \pi_1(M)$ is a monomorphism. Prove that if φ is an automorphism of $\pi_1(M)$ that preserves boundary elements of $\pi_1(M)$ ($[\alpha] \in \pi_1(M)$ is a boundary element if the loop α is freely homotopic to a loop on ∂M), then it can be realized geometrically, i.e., there exists a homeomorphism $f \colon M \to M$ inducing φ (Marden and Maskit [171]).

96. Prove that if a Kleinian group G does not contain parabolic elements and is isomorphic to the fundamental group of a closed orientable surface, then G is either quasi-Fuchsian or degenerate (Maskit [179]).

97. Prove that if a Kleinian group G does not contain parabolic elements and is isomorphic to the fundamental group of a closed nonorientable surface, then G is a $\mathbf{Z}_2$-extension of some quasi-Fuchsian group (Gusevskiĭ [92]).

98. Prove that if a finitely generated Kleinian group G does not contain elements of finite order and is a $\mathbf{Z}_2$-extension of a quasi-Fuchsian group without parabolic elements,(21) then G is isomorphic to the fundamental group of a closed nonorientable surface (Gusevskiĭ [92]).

99. Let G be a torsion-free Kleinian group satisfying the following conditions: 1) $M(G)$ is a compact manifold; 2) $\partial M(G)$ is connected; and 3) the genus of $\partial M(G)$ is equal to the number of generators of G. Prove that G is then a Schottky group (Gusevskiĭ [93]).

100. Let G be a doubly generated torsion-free Kleinian group. Prove that if the manifold $M(G)$ is compact, then G is a Schottky group (Gusevskiĭ [93]).

101. Let G be a torsion-free quasi-Fuchsian group, and let $S = \Omega_0/G$, where Ω_0 is one of the two components of G. Prove that $M(G)$ is homeomorphic to $S \times [0, 1]$ (Marden [168]).

---

(21) *Translation editor's note*: but is not itself quasi-Fuchsian,.

102. Let G be a degenerate torsion-free Kleinian group. Prove that if the surface $S = \Omega(G)/G$ is compact, then the manifold $M(G)$ is homeomorphic to $S \times [0, 1)$ (Gusevskiĭ [91]).(22)

103. Prove that if M is a compact n-dimensional manifold such that there is an infinite discontinuous group of homeomorphisms on $M \backslash \{a\}$, then M is homotopic to the sphere S^n, and is homeomorphic to S^n when $n \geq 5$ (Kulkarni [145]).

104. Prove that a connected sum of conformally Euclidean manifolds admits a conformally Euclidean structure (Kulkarni [146]).

105. Prove that if the fundamental group $\pi_1(M)$ of a three-dimensional manifold M contains an infinitely generated abelian subgroup G, then G is isomorphic to a subgroup of the additive group of rational numbers (Evans and Moser [76]).

106. Show that if the fundamental group $\pi_1(M)$ of a three-dimensional manifold M contains a finitely generated abelian subgroup G, then G is isomorphic to one of the following groups: $\mathbf{Z}, \mathbf{Z} \oplus \mathbf{Z}, \mathbf{Z} \oplus \mathbf{Z} \oplus \mathbf{Z}, \mathbf{Z}_p$ or $\mathbf{Z} \oplus \mathbf{Z}_2$ (Hempel [100]).

107. Let M be a compact irreducible orientable three-dimensional manifold, and $F \subset \partial M$ a compact surface different from a sphere or a disk. Prove that if $\pi_1(F) \to \pi_1(M)$ is an isomorphism, then M is homeomorphic to $F \times [0, 1]$ (Brown [62]).

108. Let M be a compact orientable irreducible three-dimensional manifold. If $\pi_1(M)$ contains a subgroup of finite index isomorphic to the fundamental group of a closed surface S of genus $p \geq 1$, then $\pi_1(M)$ itself is isomorphic to the fundamental group of a closed surface, and M is a line bundle (with compact fiber $I = [0, 1]$) over some closed surface (Hempel [100]).

109. Prove that if the fundamental group $\pi_1(M)$ of a three-dimensional manifold M is finitely generated, then $\pi_1(M)$ has finitely many defining relations (Scott [215]).

From this it follows, in particular, that a finitely generated Kleinian group has finitely many defining relations.

110. Prove that any three-dimensional manifold M with finitely generated fundamental group contains a compact three-dimensional submanifold M_0 such that the imbedding $i \colon M_0 \subset M$ induces an isomorphism $i_* \colon \pi_1(M_0) \to \pi_1(M)$ (Scott [216]).

111. Prove that the fundamental group of a three-dimensional compact manifold is finitely generated (Scott [215]).

112. Prove that if M is a nonorientable three-dimensional manifold with finite fundamental group, then it is homotopy equivalent to a connected sum

$$\mathbf{RP}^2 \times [0, 1] \# B^3 \# \cdots \# B^3$$

(22) *Translation editor's note*: This result, if true, is very deep; there is at this point no correct proof in the literature.

(and, in particular, $\pi_1(M) \cong \mathbf{Z}_2$) (Epstein [75]).

A manifold M is said to be *prime* if, whenever $M = M_1 \# M_2$, either M_1 or M_2 is the sphere S^n.

113. Prove that a three-dimensional prime manifold with nontrivial free fundamental group is homeomorphic to one of the following: 1) a 2-sphere bundle over the circle; 2) a solid torus; 3) a solid Klein bottle (see Exercise 131); or 4) a disk sum of solid tori and solid Klein bottles (Hempel [100]).([23])

114. Let M be a three-dimensional compact irreducible manifold. Assume that there exists an exact sequence

$$1 \to N \to \pi_1(M) \to Q \to 1, \tag{128}$$

where N is a finitely generated normal subgroup of $\pi_1(M)$ with an infinite free factor group Q.

Prove that if $Q \cong \mathbf{Z}$, then M is homeomorphic to a fiber bundle over the circle whose fiber is a two-dimensional surface; but if the rank of Q is greater than 1, then M is homeomorphic to a fiber bundle over a two-dimensional surface, with fiber the circle (Hempel [100]).

115. Let M be a three-dimensional compact manifold for which there is an exact sequence (128) with infinite factor group Q. Prove that if M is irreducible and Q is isomorphic to the fundamental group of a closed surface S (different from S^2 and $\mathbf{RP}^2$), then M is homeomorphic to a fiber bundle over S with fiber the circle (Hempel [100]).

116. Let M be a three-dimensional compact irreducible manifold. If the commutator subgroup $[\pi_1(M), \pi_1(M)]$ is finitely generated and $H_1(M, \mathbf{Z}) \cong \mathbf{Z}$, then M either is homotopy equivalent to $\mathbf{RP}^2 \times S^1$ or is a fiber bundle over the circle with fiber a compact surface S, and $[\pi_1(M), \pi_1(M)] \cong \pi_1(S)$ (Hempel [100]).

117. Let M be a three-dimensional compact irreducible orientable manifold with infinite fundamental group $\pi_1(M)$. Prove that if the center of $\pi_1(M)$ is a nontrivial finitely generated noncyclic group, then either M is homeomorphic to $S^1 \times S^1 \times S^1$, or M is a line bundle over a torus (Hempel [100]).

A three-dimensional manifold M is said to be *sufficiently large* if it contains an *incompressible two-sided surface* S (i.e., either S is a sphere not contractible in M, or the homomorphism $\pi_1(S) \to \pi_1(M)$ is a monomorphism; roughly speaking, the latter means that it is impossible to get rid of the handles on S).

118. Prove that if a three-dimensional orientable compact manifold M is irreducible and is sufficiently large, then the universal covering of M imbeds in $\mathbf{R}^3$, while the universal covering of the interior of M is homeomorphic to $\mathbf{R}^3$ (Waldhausen [240]).

([23]) *Translation editor's note*: As stated, this is not true; for example, consider the interior of a solid torus. The hypotheses should include the word "compact".

119. Prove that a three-dimensional manifold M is a line bundle (with compact fiber) over a closed nonorientable surface S of genus $p > 2$ if and only if M is homeomorphic to $M(G) = (\mathbf{R}^3_+ \cup \Omega(G))/G$, where G is a $\mathbf{Z}_2$-extension (torsion-free) of a quasi-Fuchsian group without parabolic elements (Gusevskiĭ [92],[94]).

Conclusion. Some Unsolved Problems

In conclusion we formulate some unsolved problems on Kleinian groups and uniformization of manifolds. Some of these problems are well known and are due to various authors, and some are stated for the first time. A long list of problems can also be found in, for example, the book [59], and in [303].

1. Does the limit set of each finitely generated Kleinian group in $\mathbf{R}^n$ have n-dimensional measure zero?

This problem has not yet been solved even for the two-dimensional case, though it was posed by Ahlfors as far back as 1964.

2. Is it true that if a finitely generated Kleinian group is not geometrically finite, then it contains either a degenerate subgroup or a web subgroup that cannot be constructed in the Maskit sense from elementary groups?

3. Are all degenerate and partially degenerate groups boundary groups?

4. Is there a geometric algorithm for constructing degenerate groups?

5. Give a classification of all the finitely generated Kleinian groups.

Maskit [188] classified all the finitely generated function groups.

6. Do there exist purely spatial (finitely generated) degenerate groups, i.e., groups not obtainable by trivial extension of planar groups to space or by quasiconformal deformations of them?

7. Are the assertions in Problem 14 true for arbitrary finitely generated Kleinian groups?

8. Can a Fuchsian group $G \subset \mathcal{M}_n$ of the second kind satisfying condition (63) have a limit set of positive $(n-1)$-dimensional measure?

For $n = 2$ this is possible, as shown by results of Golubev [85]. However, these results make essential use of properties of analytic functions on the plane.

9. Give a geometric proof of the fact that a finitely generated Kleinian group acting in $\overline{\mathbf{R}}^n$ $(n \geq 2)$ has a subgroup of finite index not containing elliptic elements.

An algebraic proof of this fact follows from results of Mal'tsev [286] on matrix groups (1940); another proof based on number-theoretic ideas was obtained by Selberg [219].

10. For what discrete groups $\Gamma \subset G \subset \mathcal{M}_n$ are the groups $B(\Gamma)$ and $C(\Gamma)$ in Exercise 38 discrete?

11. Let $\Gamma \subset \mathcal{M}_n$ be a discrete nonarithmetic subgroup (see [40]) with $\mathrm{vol}(\mathcal{M}_n/\Gamma) < \infty$. Can the subgroup $B(\Gamma)$ be dense in $\mathcal{M}_n$?

If Γ is a discrete arithmetic group and $\mathrm{vol}(\mathcal{M}_n/\Gamma) < \infty$, then the subgroup is dense.

12. Let G be a nonconstructible web group. Can the manifold $M(G)$ be imbedded in a compact manifold with boundary?

13. Describe the variety $\mathrm{Hom}_a(\Gamma, \mathcal{M}_2)$. Is a neighborhood of the identity in it a cell?

14. Are there necessary and sufficient conditions for stability?

15. Are quasiconformal deformations of stable groups also stable?

16. Are there stable Kleinian groups that are not geometrically finite?

17. Give a constructive (explicit) description of the kernel of the Poincaré Θ-operator (see §5 in Chapter I).

Here there is the result of Lyan [155], [156] determining the functions forming the kernel. However, an explicit form has not been established for them. Kra [279] has made further essential progress.

18. Let S and S' be closed Riemann surfaces of genera p and p', respectively ($p \geq p' > 1$). Determine the number of possible conformal mappings from S to S'.

19. Determine the number of nonequivalent n-sheeted coverings over a closed Riemann surface of genus p with given orders at the branch points.

The results for unbranched coverings is contained in Problem 77.

20. Describe all the closed three-dimensional manifolds admitting a complete hyperbolic structure.

21. Which three-dimensional manifolds M with boundary can be uniformized by Kleinian groups, i.e., can be represented in the form $M = M(G) = (\mathbf{R}^3_+ \cup \Omega)/G$?

22. Describe all the links (with finitely many components) whose complements admit complete hyperbolic structures.

There are some results in this direction, for example, in [211]. A complete solution of the problem for knots has been announced by Thurston (see §7.2 in Chapter II).

23. Do there exist nontrivial quasiconformal deformations of a conformal structure on a compact hyperbolic manifold that is uniformizable by a Fuchsian group $G \subset \mathcal{M}_3$ having a simplex as fundamental polyhedron?

24. As noted, the (Teichmüller) space $T(G)$ of quasiconformal deformations of each Kleinian group $G \subset \mathcal{M}_2$ is a cell of definite dimension. What is the structure of the space of quasiconformal deformations of a Fuchsian group $G \subset \mathcal{M}_n$? Is it a cell, and, if so, what is the dimension?

25. Let $G \subset \mathcal{M}_n$ be a finitely generated Fuchsian group of the first kind. For $n \geq 3$ does the boundary of the space $T(G)$ contain Kleinian groups with a single invariant component, i.e., *b*-groups?

The results obtained in [36] and [37] on the boundary of a Teichmüller space and the arguments related to them permit us to conjecture that there are no such boundary groups in space.

26. A manifold M^n (n-dimensional) is said to be *homologically flat* if each $(n-1)$-dimensional sphere imbedded in it separates M^n into two components. Describe all the normal subgroups of $\pi_1(M^n)$ that serve as defining subgroups of homologically flat coverings of M^n.

27. A manifold M^n is said to be *flat* if it is homeomorphic to some region in $\overline{\mathbf{R}}^n$. Find necessary and sufficient conditions for a homologically flat manifold to be flat. (This is always so for $n = 2$, but for $n > 2$ there are examples of homologically flat manifolds that are not flat.)

28. Is every homologically flat infinite-sheeted covering of a closed manifold flat?

29. Let M_1 and M_2 be two n-dimensional homotopy equivalent manifolds without boundary, one of which can be topologically uniformized by some Kleinian group in $\overline{\mathbf{R}}^n$. Can the second manifold then also be uniformized by a Kleinian group?

30. Does each three-dimensional manifold with finite fundamental group have an elliptic structure?

31. Let M be a closed orientable aspherical three-dimensional manifold. Does there exist a finite-sheeted covering $\tilde{M}$ of M such that $H_1(M, Z)$ is infinite? Does there exist a finite-sheeted covering $\tilde{M}$ of M that is sufficiently large?

32. Let M be a closed three-dimensional hyperbolic manifold, and Γ a finite group of diffeomorphisms of M. Is Γ conjugate to a group of isometries of M in the whole diffeomorphism group of M?

33. Let M be a closed three-dimensional hyperbolic manifold, and f: $M \to M$ a diffeomorphism of finite order. Is f equivariantly isotopic to an isometry? Is the set of fixed points of f a union of geodesics?

34. Let $M = S_g \times S^1$, where S_g is a closed surface of genus $g \geq 2$, and S^1 is the circle. Do there exist conformal structures on M for which a development is a mapping onto the whole of $\overline{\mathbf{R}}^3$ (see §8 in Chapter II)?

35. Describe all Seifert manifolds on which there exist conformal structures.

36. Describe the closed three-dimensional manifolds with conformal structures.

37. Do there exist conformally rigid conformal manifolds? (A conformal manifold M is said to be *conformally rigid* if any conformal manifold N diffeomorphic to M is conformally equivalent to M.)

38. Let M be a closed irreducible three-dimensional manifold homotopically equivalent to a hyperbolic manifold. Is M itself hyperbolic?

39. Suppose that a three-dimensional M has a finite-sheeted covering by a closed hyperbolic manifold. Is M itself hyperbolic?

40. Is a torus sum of two three-dimensional conformal manifolds a conformal manifold? A *torus sum* of two three-dimensional manifolds M_1 and M_2 is defined to be a manifold obtained by gluing together the boundaries of solid tori removed from M_1 and M_2, by means of some diffeomorphism.

41. Suppose that a three-dimensional manifold M has a finite-sheeted covering by a conformal manifold. Is M itself conformal?

42. Let M_f^g be a manifold obtained from the shell $S_g \times [0,1]$ (S_g is a closed orientable surface of genus $g \geq 2$) by identifying $S_g \times \{0\}$ and $S_g \times \{1\}$ by means of a diffeomorphism $f\colon S_g \to S_g$. Describe the classes of diffeomorphisms f such that M_f^g has a conformal structure.

REMARK. Since publication of the Russian edition, Problems 11, 14–16, 18, and 19 have been solved. Specifically, 11 was solved by Margulis (see Problem 40 in §3 of Chapter IV), the solutions of 14–16 follow from results of Sullivan (see the end of §5 in Chapter I), Problem 19 was solved by Mednykh [292], and as for Problem 18, see Problem 60B).

Bibliography

[**1**] Herbert Abels, *Geometrische Erzeugung von diskontinuierlichen Gruppen*, Schriftenreihe Math. Inst. Univ. Münster No. 33 (1966).

[**2**] William Abikoff, *Some remarks on Kleinian groups*, Advances in the Theory of Riemann Surfaces (Proc. Conf., Stony Brook, N. Y., 1969; L. V. Ahlfors et al., editors), Ann. of Math. Studies, vol. 66, Princeton Univ. Press, Princeton, N. J., and Univ. of Tokyo Press, Tokyo, 1971, pp. 1–5.

[**3**] ——, *Remarks on a paper of Maskit*, Preprint, Columbia Univ., New York, 1973.

[**4**] ——, *The residual limit sets of Kleinian groups*, Acta Math. **130** (1973), 127–144.

[**5**] ——, *Constructability and Bers stability of Kleinian groups*, Discontinuous Groups and Riemann Surfaces (Proc Conf., University Park, Md., 1973; L. Greenberg, editor), Ann. of Math. Studies, vol. 79, Princeton Univ. Press, Princeton, N. J., and Univ. of Tokyo Press, Tokyo, 1974, pp. 3–12.

[**6**] ——, *On boundaries of Teichmüller spaces and on Kleinian groups.* III, Acta Math. **134** (1975), 211–237.

[**7**] ——, *Degenerating families of Riemann surfaces*, Ann. of Math. (2) **105** (1977), 29–44.

[**8**] William Abikoff and Bernard Maskit, *Geometric decompositions of Kleinian groups*, Amer. J. Math. **99** (1977), 687–697.

[**9**] Robert D. M. Accola, *Invariant domains for Kleinian groups*, Amer. J. Math. **88** (1966), 329–334.

[**10**] ——, *Automorphisms of Riemann surfaces*, J. Analyse Math. **18** (1967), 1–5.

[**11**] ——, *On the number of automorphisms of a closed Riemann surface*, Trans. Amer. Math. Soc. **131** (1968), 398–408.

[**12**] ——, *Strongly branched coverings of closed Riemann surfaces*, Proc. Amer. Math. Soc. **26** (1970), 315–322.

[**13**] Lars V. Ahlfors, *On quasiconformal mappings*, J. Analyse Math. **3** (1953/54), 1–58, 207–208.

[**14**] ——, *Curvature properties of Teichmüller's space*, J. Analyse Math. 9 (1961), 161–176.

[**15**] ——, *Finitely generated Kleinian groups*, Amer. J. Math. **86** (1964), 413–429; **87** (1965), 759.

[**16**] ——, *Fundamental polyhedrons and limit point sets of Kleinian groups*, Proc. Nat. Acad. Sci. U.S.A. **55** (1966), 251–254.

[**17**] ——, *Kleinsche Gruppen in der Ebene und im Raum*, Festband zum 70. Geburtstag von Rolf Nevanlinna (H. P. Kunzi and A. Pfluger, editors), Springer-Verlag, 1966, pp. 7–15.

[**18**] ——, *Eichler integrals and the area theorem of Bers*, Michigan Math. J. **15** (1968), 257–263.

[**19**] ——, *Lectures on quasiconformal mappings*, Van Nostrand, 1966.

[**20**] ——, *The structure of a finitely generated Kleinian group*, Acta Math. **122** (1969), 1–17.

[**21**] ——, *Two lectures on Kleinian groups*, Proc. Romanian-Finnish Sem. Teichmüller Spaces and Quasiconformal Mappings (Braşov, 1969; C. A. Cazacu, editor), Publ. House Acad. Socialist Republic of Romania, Bucharest, 1971, pp. 49–64.

[**22**] Lars V. Ahlfors and Leo Sario, *Riemann surfaces*, Princeton Univ. Press, Princeton, N. J., 1960.

[**23**] L. Ahlfors and G. Weill, *A uniqueness theorem for Beltrami equations*, Proc. Amer. Math. Soc. **13** (1962), 975–978.

[**24**] Tohru Akaza, *Poincaré theta series and singular sets of Schottky groups*, Nagoya Math. J. **24** (1964), 43–65.

[**25**] A. D. Aleksandrov, *On the filling of space by polyhedra*, Vestnik Leningrad. Univ. **1954**, no. 2 (Ser. Mat. Fiz. Khim. vyp. 1), 33–43. (Russian)

[**26**] B. N. Apanasov, *Some characteristics of conformal mappings in* R^n, Metric Questions of the Theory of Functions and Mappings, No. 5, "Naukova Dumka", Kiev, 1974, pp. 3–13. (Russian)

[**27**] ——, *On a class of Kleinian groups in* R^n, Dokl. Akad. Nauk SSSR **215** (1974), 509–510=Soviet Math. Dokl. **15** (174), 515–517.

[**28**] ——, *On an analytic method in the theory of Kleinian groups on a multidimensional Euclidean space*, Dokl. Akad. Nauk SSSR **222** (1975), 11–14=Soviet Math. Dokl. **16** (1975), 553–556.

[**29**] ——, *Kleinian groups in space*, Sibirsk. Mat. Zh. **16** (1975), 891–898=Siberian Math. J. **16** (1975), 679–684.

[**30**] ——, *A universal property of Kleinian groups in the hyperbolic metric*, Dokl. Akad. Nauk SSSR **225** (1975), 15–17=Soviet Math. Dokl. **16** (1975), 1418–1421.

[**31**] ——, *On the relation of the group* $SL(n, \mathbf{Z})$ *with the* n*-dimensional modular group*, Dokl. Akad. Nauk SSSR **231** (1976), 1033–1036=Soviet Math. Dokl. **17** (1976), 1670–1674.

[**32**] ——, *On an automorphic mapping in space*, Some Questions of Contemporary Function Theory (Conf. Materials), Inst. Mat. Sibirsk. Otdel. Akad. Nauk SSSR, Novosibirsk, 1976. (Russian)

[**33**] ——, *On Mostow's rigidity theorem*, Dokl. Akad. Nauk SSSR **243** (1978), 829–832=Soviet Math. Dokl. **19** (1978), 1408–1412.

[34] ——, *Entire automorphic forms in* R^n, Sibirsk. Mat. Zh. **19** (1978), 735–748=Siberian Math. J. **19** (1978), 518–528.

[35] ——, *On the problem of the existence of automorphic mappings in n dimensions*, Questions in the Metric Theory of Mappings and Its Application (Proc. Fifth Colloq., Donetsk, 1976; G. D. Suvorov and V. I. Biliĭ, editors), "Naukova Dumka", Kiev, 1978, pp. 3–9. (Russian)

[36] ——, *Nontriviality of Teichmüller space for Kleinian group in space*, Riemann Surfaces and Related Topics: Proc. 1978 Stony Brook Conf. (I. Kra and B. Maskit, editors), Ann. of Math. Studies, vol. 97, Princeton Univ. Press, Princeton, N. J., and Univ. of Tokyo Press, Tokyo, 1981, pp. 21–31.

[37] ——, *Kleinian groups, Teichmüller space, and Mostow's rigidity theorem*, Sibirsk. Mat. Zh. **21** (1980), no. 4, 3–15=Siberian Math. J. **21** (1980), 483–491.

[38] B. N. Apanasov and A. V. Tetenov, *On the existence of nontrivial quasiconformal deformations of Kleinian groups in space*, Dokl. Akad. Nauk SSSR **239** (1978), 14–17=Soviet Math. Dokl. **19** (1978), 242–245.

[39] Paul Appell and Édouard Goursat, *Théorie des fonctions d'une variable et des transcendantes qui s'y rattachent.* Vol. II: *Fonctions automorphes*, 2nd ed. (rev. by P. Fatou), Gauthier-Villars, Paris, 1930.

[40] *Arithmetic groups and automorphic functions*, "Mir", Moscow, 1979. (Russian) This is an abridged Russian translation of the following three items:

a. *Algebraic groups and discontinuous subgroups*, Proc. Sympos. Pure Math., vol. 9, Amer. Math. Soc., Providence, R. I., 1966.

b. Roger Godement, *Introduction à la théorie de Langlands*, Sém. Bourbaki 1966/67, Exposé 321, Secrétariat Math., Paris, 1967; reprint, Benjamin, 1968.

c. Atle Selberg, *Discrete groups and harmonic analysis*, Proc. Internat. Congr. Math. (Stockholm, 1962), Inst. Mittag-Leffler, Djursholm, 1963, pp. 177–189.

[41] Walter L. Baily, Jr., *Introductory lectures on automorphic forms*, Publ. Math. Soc. Japan, vol. 12, Iwanami Shoten, Tokyo, and Princeton Univ. Press, Princeton, N. J., 1973.

[42] A. F. Beardon, *The Hausdorff dimension of singular sets of properly discontinuous groups*, Amer. J. Math. **88** (1966), 722–736.

[43] Alan F. Beardon and Troels Jørgensen, *Fundamental domains for finitely generated Kleinian groups*, Math. Scand. **36** (1975), 21–26.

[44] Alan F. Beardon and Bernard Maskit, *Limit points of Kleinian groups and finite sided fundamental polyhedra*, Acta Math. **132** (1974), 1–12.

[45] P. P. Belinskiĭ, *General properties of quasiconformal mappings*, "Nauka", Novosibirsk, 1974. (Russian)

[46] ——, *An example of a universal Riemann surface containing any compact surface with boundary as a subregion*, Sibirsk. Mat. Zh. **21** (1980), no. 3, 224–225. (Russian)

[**47**] Lipman Bers, *Quasiconformal mappings and Teichmüller's theorem*, Analytic Functions (Conf., Princeton, N. J., 1957; by R. Nevanlinna et al.), Princeton Univ. Press, Princeton, N. J., 1960, pp. 89–119.

[**48**] ——, *On moduli of Riemann surfaces*, Lectures, Forschungsinst. Math., E.T.H., Zürich, 1964.

[**49**] ——, *Inequalities for finitely generated Kleinian groups*, J. Analyse Math. **18** (1967), 23–41.

[**50**] ——, *Spaces of Kleinian groups*, Several Complex Variables (Proc. Conf., College Park, Md., 1970), Lecture Notes in Math., vol. 155, Springer-Verlag, 1970, pp. 9–34.

[**51**] ——, *On boundaries of Teichmüller spaces and on Kleinian groups. I*, Ann. of Math. (2) **91** (1970), 570–600.

[**52**] ——, *A non-standard integral equation with applications to quasiconformal mappings*, Acta Math. **116** (1966), 113–134.

[**53**] ——, *A remark on Mumford's compactness theorem*, Israel J. Math. **12** (1972), 400–407.

[**54**] ——, *Uniformization, moduli, and Kleinian groups*, Bull. London Math. Soc. **4** (1972), 257–300.

[**55**] ——, *On moduli of Kleinian groups*, Uspekhi Mat. Nauk **29** (1974), no. 2 (1976), 86–102=Russian Math. Surveys **29** (1974), no. 2, 88–102.

[**56**] ——, *Deformations and moduli of Riemann surfaces with nodes and signatures*, Math. Scand. **36** (1975), 12–16.

[**57**] ——, *On Hilbert's 22nd problem*, Mathematical Developments Arising from Hilbert Problems, Proc. Sympos. Pure Math., vol. 28, Amer. Math. Soc., Providence, R. I., 1976, pp. 559–609.

[**58**] Lipman Bers and Leon Ehrenpreis, *Holomorphic convexity of Teichmüller spaces*, Bull. Amer. Math. Soc. **70** (1964), 761–764.

[**59**] L. Bers and I. Kra (editors), *A crash course on Kleinian groups* (Lectures, San Francisco, 1974), Lecture Notes in Math., vol. 400, Springer-Verlag, 1974.

[**60**] A. Beurling and L. Ahlfors, *The boundary correspondence under quasiconformal mappings*, Acta Math. **96** (1956), 125–142.

[**61**] Joan S. Birman, *Abelian quotients of the mapping class group of a 2-manifold*, Bull. Amer. Math. Soc. **76** (1970), 147–150.

[**62**] E. M. Brown, *Unknotting in $M^2 \times I$*, Trans. Amer. Math. Soc. **123** (1966), 480–505.

[**63**] Henri Cartan, *Quotient d'un espace analytique par un groupe d'automorphismes*, Algebraic Geometry and Topology (Sympos. in Honor of S. Lefschetz, Princeton, N. J., 1954; R. Fox, D. Spencer and A. Tucker, editors), Princeton Univ. Press, Princeton, N. J., 1957, pp. 90–102.

[**64**] N. G. Chebotarev, *Theory of algebraic functions*, GITTL, Moscow, 1948. (Russian)

[**65**] Su-Shing Chen, *A theorem of Ahlfors for hyperbolic spaces*, Trans. Amer. Math. Soc. **242** (1978), 401–406.

[**66**] Vicki Chuckrow, *On Schottky groups with applications to Kleinian groups*, Ann. of Math. (2) **88** (1968), 47–61.

[**67**] V. V. Chueshev, *Schottky spaces of the type* (g, s, m), Dokl. Akad. Nauk SSSR **223** (1975), 1326–1328=Soviet Math. Dokl. **16** (1975), 1107–1110.

[**68**] ——, *Spaces of Kleinian groups and Riemann surfaces*, Candidate's Dissertation, Inst. Mat. Sibirsk. Otdel. Akad. Nauk SSSR, Novosibirsk, 1979. (Russian)

[**69**] Richard H. Crowell and Ralph H. Fox, *Introduction to knot theory*, Ginn, Boston, Mass., 1963.

[**70**] Clifford J. Earle, *On holomorphic cross-sections in Teichmüller spaces*, Duke Math. J. **36** (1969), 409–415.

[**71**] Clifford J. Earle and James Eells, *A fibre bundle description of Teichmüller theory*, J. Differential Geometry **3** (1969), 19–43.

[**72**] Clifford J. Earle and Irwin Kra, *On holomorphic mappings between Teichmüller spaces*, Contributions to Analysis (Collection Dedicated to Lipman Bers; L. Ahlfors et al., editors), Academic Press, 1974, pp. 107–124.

[**73**] ——, *On sections of some holomorphic families of closed Riemann surfaces*, Acta Math. **137** (1976), 49–79.

[**74**] D. B. A. Epstein, *Finite presentations of groups and 3-manifolds*, Quart J. Math. Oxford Ser. (2) **12** (1961), 205–212.

[**75**] ——, *Projective planes in* 3*-manifolds*, Proc. London Math. Soc. (3) **11** (1961), 469–484.

[**76**] Benny Evans and Louise Moser, *Solvable fundamental groups of compact* 3*-manifolds*, Trans. Amer. Math. Soc. **168** (1972), 189–210.

[**77**] W. Fenchel and J. Nielsen, *On discontinuous groups of isometric transformations of the non-Euclidean plane*, Studies and Essays Presented to R. Courant on his 60th Birthday (K. Friedrichs, O. Neugebauer and J. Stoker, editors), Interscience, 1948, pp. 117–128.

[**78**] L. R. Ford, *Automorphic functions*, McGraw-Hill, 1929.

[**79**] Robert Fricke and Felix Klein, *Vorlesungen über die Theorie der automorphen Functionen.* Vols. I, II, 2nd ed., Teubner, Leipzig, 1926; reprint, 1965.

[**80**] F. Gardiner and I. Kra, *Stability of Kleinian groups*, Indiana Univ. Math. J. **21** (1971/72), 1037–1059.

[**81**] H. Garland and M. S. Raghunathan, *Fundamental domains for lattices in (R-) rank* 1 *semisimple Lie groups,* Ann. of Math. (2) **92** (1970), 279–326.

[**82**] F. W. Gehring, *Spirals and the universal Teichmüller space*, Acta Math. **141** (1978), 99–113.

[**83**] ——, *Univalent functions and the Schwarzian derivative*, Comment. Math. Helv. **52** (1977), 561–572.

[**84**] V. V. Golubev, *Lectures on the analytic theory of differential equations*, 2nd ed., GITTL, Moscow, 1950. (Russian)

[**85**] ——, *Single-valued analytic functions. Automorphic functions*, Fizmatgiz, Moscow, 1961. (Russian)

[**86**] Leon Greenberg, *Discrete subgroups of the Lorentz group*, Math. Scand. **10** (1962), 85–107.

[**87**] ——, *Fundamental polyhedra for Kleinian groups*, Ann. of Math. (2) **84** (1966), 433–441.

[**88**] ——, *Commensurable groups of Moebius transformations*, Discontinuous Groups and Riemann Surfaces (Proc. Conf., University Park, Md., 1973; L. Greenberg, editor), Ann. of Math. Studies, vol. 79, Princeton Univ. Press, Princeton, N. J., and Univ. of Tokyo Press, Tokyo, 1974, pp. 227–237.

[**89**] ——, *Maximal groups and signatures*, Discontinuous Groups and Riemann Surfaces (Proc. Conf., University Park, Md., 1973; L. Greenberg, editor), Ann. of Math. Studies, vol. 79, Princeton Univ. Press, Princeton, N. J., and Univ. of Tokyo Press, Tokyo, 1974, pp. 207–226.

[**90**] Phillip A. Griffiths, *Complex-analytic properties of certain Zariski-open sets on algebraic varieties*, Ann. of Math. (2) **94** (1971), 21–51.

[**91**] N. A. Gusevskiĭ, *On a problem in the theory of Kleinian groups*, Dokl. Akad. Nauk SSSR **234** (1977), 277–279=Soviet Math. Dokl. **18** (1977), 629–631.

[**92**] ——, *On two classes of Kleinian groups*, Dokl. Akad. Nauk SSSR **241** (1978), 753–756=Soviet Math. Dokl. **19** (1978), 919–922.

[**93**] ——, *Topological characterization of Schottky groups*, Sibirsk. Mat. Zh. **20** (1979), 651–655=Siberian Math. J. **20** (1979), 454–457.

[**94**] ——, *Kleinian groups and 3-manifolds*, Candidate's Dissertation, Inst. Mat. Sibirsk. Otdel. Akad. Nauk SSSR, Novosibirsk, 1979. (Russian)

[**95**] V. Ya. Gutlyanskiĭ, *On the area principle for a class of quasiconformal mappings*, Dokl. Akad. Nauk SSSR **212** (1973), 540–543=Soviet Math. Dokl. **14** (1973), 1401–1406.

[**96**] Zh. Adamar [Jacques Hadamard], *Non-Euclidean geometry in the theory of automorphic functions*, GITTL, Moscow, 1951. (Russian)

[**97**] W. J. Harvey (editor), *Discrete groups and automorphic functions* (Proc. Instructional Conf., Cambridge, 1975), Academic Press, 1977.

[**98**] Dennis A. Hejhal, *On Schottky and Teichmüller spaces*, Advances in Math. **15** (1975), 133–156.

[**99**] ——, *The variational theory of linearly polymorphic functions,* J. Analyse Math. **30** (1976), 215–264.

[**100**] John Hempel, 3-*manifolds*, Ann. of Math. Studies, vol. 86, Princeton Univ. Press, Princeton, N. J., and Univ. of Tokyo Press, Tokyo, 1976.

[**101**] John Hempel and William Jaco, 3-*manifolds which fiber over a surface,* Amer. J. Math. **94** (1972), 189–205.

[**102**] ____, *Fundamental groups of 3-manifolds which are extensions*, Ann. of Math. (2) **95** (1972), 86–98.

[**103**] John Hamal Hubbard, *Sur les sections analytiques de la courbe universelle de Teichmüller*, Mem. Amer. Math. Soc. **4** (1976), no. 166.

[**104**] A. Hurwitz, *Über algebraische Gebilde mit eindeutigen Transformationen in sich*, Math. Ann. **41** (1893), 403–442.

[**105**] Yu. G. Il′yashenko, *Nondegenerate B-groups*, Dokl. Akad. Nauk SSSR **208** (1973), 1020–1022=Soviet Math. Dokl. **14** (1973), 207–210.

[**106**] ____, *Nondegenerate B-groups and simultaneous uniformization*, Some Questions of Contemporary Function Theory (Conf. Materials, Novosibirsk, 1976; P. P. Belinskiĭ, editor), Inst. Mat. Sibirsk. Otdel. Akad. Nauk SSSR, Novosibirsk, 1976, pp 67–74. (Russian)

[**107**] William Jaco, *Finitely presented subgroups of three-manifold groups*, Invent. Math. **13** (1971), 335–346.

[**108**] Troels Jørgensen, *Some remarks on Kleinian groups*, Math. Scand. **34** (1974), 101–108.

[**109**] ____, *On discrete groups of Möbius transformations*, Amer. J. Math. **98** (1976), 739–749.

[**110**] ____, *Compact 3-manifolds of constant negative curvature fibering over the circle*, Ann. of Math. (2) **106** (1977), 61–72.

[**111**] Troels Jørgensen and Maire Kiikka, *Some extreme discrete groups*, Ann. Acad. Sci. Fenn. Ser. A I Math. **1** (1975), 245–248.

[**112**] Troels Jørgensen and Albert Marden, *Two doubly degenerate groups*, Quart. J. Math. Oxford Ser. (2) **30** (1979), 143–156.

[**113**] T. Jørgensen, A. Marden and B. Maskit, *The boundary of classical Schottky space*, Duke Math. J. **46** (1979), 441–446.

[**114**] Linda Keen, *Canonical polygons for finitely generated Fuchsian groups*, Acta Math. **115** (1965), 1–16.

[**115**] ____, *Intrinsic moduli on Riemann surfaces*, Ann. of Math. (2) **84** (1966), 404–420.

[**116**] ____, *A rough fundamental domain for Teichmüller spaces*, Bull. Amer. Math. Soc. **83** (1977), 1199–1226.

[**117**] Felix Klein, *Neue Beiträge zur Riemann'schen Functionentheorie,* Math. Ann. **21** (1883), 141–218.

[**118**] ____, *Gesammelte mathematische Abhandlungen.* Vol. 3, Springer-Verlag, 1923.

[**119**] Paul Koebe, *Über die Uniformisierung der algebraischen Kurven.* I, II, III, IV, Math. Ann. **67** (1909), 145–224; **69** (1910), 1–81; **72** (1912), 437–516; **75** (1914), 42–129.

[**120**] Irwin Kra, *On spaces of Kleinian groups*, Comment. Math. Helv. **47** (1972), 53–69.

[**121**] ____, *Automorphic forms and Kleinian groups*, Benjamin, 1972.

[**122**] ——, *On new kinds of Teichmüller spaces*, Israel J. Math. **16** (1973), 237–257.

[**123**] ——, *Varieties of Kleinian groups*, Complex Analysis 1972, Vol. II (Proc. Conf., Houston, Texas; H. Resnikoff and R. Wells, editors), Rice Univ. Studies **59** (1973), no. 2, 41–56.

[**124**] I. Kra and B. Maskit, *Involutions on Kleinian groups*, Bull. Amer. Math. Soc. **78** (1972), 801–805.

[**125**] Wilhelm Kraus, *Über die Zusammenhang einiger Charakteristiken eines einfach zusammenhängenden Bereiches mit der Kreisabbildung*, Mitt. Math. Sem. Univ. Giessen **21** (1932).

[**126**] S. N. Krushkal′, *On the theory of extremal problems for quasiconformal mappings of closed Riemann surfaces*, Dokl. Akad. Nauk SSSR **171** (1966), 784–787=Soviet Math. Dokl **7** (1966), 1541–1544.

[**127**] ——, *Teichmüller's theorem on extremal quasiconformal mappings*. Sibirsk. Mat. Zh. **8** (1967), 313–332=Siberian Math. J. **8** (1967), 231–244.

[**128**] ——, *Moduli of Riemann surfaces*, Sibirsk. Mat. Zh. **13** (1972) 349–367=Siberian Math. J. **13** (1972), 241–253.

[**129**] ——, *Variational methods in the theory of quasiconformal mappings* Novosibirsk. Gos. Univ., Novosibirsk, 1974. (Russian)

[**130**] ——, *Quasiconformal mappings and Riemann surfaces*, "Nauka" Novosibirsk, 1975. (Russian; English transl. listed separately as [137])

[**131**] ——, *On a property of limit sets of Kleinian groups*, Dokl Akad. Nauk SSSR **225** (1975), 500–502; **237** (1977), 256=Soviet Math. Dokl. **16** (1975), 1497–1499; **18** (1977), no. 4, iv.

[**132**] ——, *On the supports of Beltrami differentials for Kleinian groups*, Dokl. Akad. Nauk SSSR **231** (1976), 799–801=Soviet Math. Dokl. **17** (1976), 1642–1644.

[**133**] ——, *Two theorems on Teichmüller spaces*, Dokl. Akad. Nauk SSSR **228** (1976), 290–292=Soviet Math. Dokl. **17** (1976), 704–707.

[**134**] ——, *Some remarks on Kleinian groups*, Riemann Surfaces and Related Topics: Proc. 1978 Stony Brook Conf. (I. Kra and B. Maskit, editors), Ann. of Math. Studies, vol. 97, Princeton Univ. Press, Princeton, N. J., and Univ. of Tokyo Press, Tokyo, 1981, pp. 361–366.

[**135**] ——, *Some rigidity theorems for discontinuous groups*, Math. Anal. and Related Math. Questions (A. A. Borovkov, editor), "Nauka", Novosibirsk, 1978, pp. 69–82; English transl., Amer. Math. Soc. Transl. (2) **122** (1984), 75–83.

[**136**] ——, *Quasiconformal stability of Kleinian groups*, Sibirsk. Mat. Zh. **20** (1979), 322–329=Siberian Math. J. **20** (1979), 229–234.

[**137**] ——, *Quasiconformal mappings and Riemann surfaces*, Wiley, 1979.

[**138**] ——, *To the problem of the supports of Beltrami differentials for Kleinian groups*, Romanian-Finnish Sem. Complex Anal. (Proc., Bucharest, 1976), Lecture Notes in Math., vol. 743, Springer-Verlag, 1979, pp. 132–139.

[**139**] ———, *A remark on rigidity of quasiconformal deformations of discrete isometry groups of hyperbolic spaces*, Sibirsk. Mat. Zh. **21** (1980), no. 5, 52–57=Siberian Math. J. **21** (1980), 683–687.

[**140**] S. L. Krushkal′, B. N. Apanasov and N. A. Gusevskiĭ, *Kleinian groups in examples and problems*, Novosibirsk. Gos. Univ., Novosibirsk, 1978. (Russian)

[**141**] ———, *Uniformization and Kleinian groups*, Novosibirsk. Gos. Univ., Novosibirsk, 1979. (Russian)

[**142**] R. Kühnau, *Geometrie der konformen Abbildung auf der hyperbolischen und der elliptischen Ebene*, VEB Deutscher Verlag Wiss., Berlin, 1974.

[**143**] N. H. Kuiper, *On compact conformally Euclidean spaces of dimension* > 2, Ann. of Math. (2) **52** (1950), 478–490.

[**144**] ———, *On conformally-flat spaces in the large*, Ann. of Math. (2) **50** (1949), 916–924.

[**145**] R. S. Kulkarni, *Infinite regular coverings*, Duke Math. J. **45** (1978), 781–796.

[**146**] ———, *On the principle of uniformization*, J. Differential Geometry **13** (1978), 109–138.

[**147**] Herbert Lange, *Moduli spaces of algebraic curves with rational maps*, Math. Proc. Cambridge Philos. Soc. **78** (1975), 283–292.

[**148**] M. A. Lavrent′ev, *Variational methods for boundary value problems for systems of elliptic equations*, Izdat. Akad. Nauk SSSR, Moscow, 1962; English transl., Noordhoff, 1963.

[**149**] Joseph Lehner, *Discontinuous groups and automorphic functions*, Math. Surveys, Vol. 8, Amer. Math. Soc., Providence, R. I., 1964; reprinted with corrections, 1982.

[**150**] Olli Lehto, *Univalent functions and Teichmüller theory*, Proc. First Finnish-Polish Summer School Complex Anal. (Podlesice, 1977; J. Ławrynowicz and O. Lehto, editors), Part 1, Univ. Łódź, Łódź, 1978, pp. 11–33.

[**151**] O. Lehto and K. I. Virtanen, *Quasikonforme Abbildungen*, Springer-Verlag, 1965.

[**152**] Armin Leutbecher, *Über die Heckeschen Gruppen* $\mathfrak{G}(\lambda)$, Abh. Math. Sem. Univ. Hamburg **31** (1967), 199–205.

[**153**] W. B. R. Lickorish, *A finite set of generators of the homeotopy group of a 2-manifold*, Proc. Cambridge Philos. Soc. **60** (1964), 769–778; **62** (1966), 679–681.

[**154**] Michele Linch, *A comparison of metrics on Teichmüller space*, Proc. Amer. Math. Soc. **43** (1974), 349–352.

[**155**] G. M. Lyan, *On the kernel of Poincaré's* θ*-operator*, Dokl. Akad. Nauk SSSR **230** (1976), 269–270=Soviet Math. Dokl. **17** (1976), 1283–1285.

[**156**] ———, *A constructive description of the kernel of the Poincaré* θ*-operator*, Dokl. Akad. Nauk UzSSR **1977**, no. 12, 7–8. (Russian)

[**157**] ——, *Teichmüller space and the Poincaré Θ-operator*, Dokl. Akad. Nauk SSSR **250** (1980), 1053–1056=Soviet Math. Dokl. **21** (1980), 260–262.

[**158**] A. M. Macbeath, *On a theorem of Hurwitz*, Proc. Glasgow Math. Assoc. **5** (1961/62), 90–96.

[**159**] ——, *On a curve of genus* 7, Proc. London Math. Soc. (3) **15** (1965), 527–542.

[**160**] Colin Maclachlan, *A bound for the number of automorphisms of a compact Riemann surface*, J. London Math. Soc. **44** (1969), 265–272.

[**161**] ——, *Modulus space is simply-connected*, Proc. Amer. Math. Soc. **29** (1971), 85–86.

[**162**] ——, *Maximal normal Fuchsian groups*, Illinois J. Math. **15** (1971), 104–113.

[**163**] C. Maclachlan and W. J. Harvey, *On mapping-class groups and Teichmüller spaces*, Proc. London Math. Soc. (3) **30** (1975), 496–512.

[**164**] Wilhelm Magnus, Abraham Karrass and Donald Solitar, *Combinatorial group theory*, Interscience, 1966.

[**165**] V. S. Makarov, *A certain class of discrete Lobachevsky space groups with an infinite fundamental region of finite measure*, Dokl. Akad. Nauk SSSR **167** (1966), 30–33=Soviet Math. Dokl. **7** (1966), 328–331.

[**166**] Albert Marden, *On homotopic mappings of Riemann surfaces*, Ann. of Math. (2) **90** (1969), 1–8.

[**167**] ——, *An inequality for Kleinian groups*, Advances in the Theory of Riemann Surfaces (Proc. Conf., Stony Brook, N. Y., 1969; L. V. Ahlfors et al., editors), Ann. of Math. Studies, vol. 66, Princeton Univ. Press, Princeton, N. J., and Univ. of Tokyo Press, Tokyo, 1971, pp. 295–296.

[**168**] ——, *The geometry of finitely generated Kleinian groups*, Ann. of Math. (2) **99** (1974), 383–462.

[**169**] ——, *Universal properties of Fuchsian groups in the Poincaré metric*, Discontinuous Groups and Riemann Surfaces (Proc. Conf., University Park, Md., 1973; L. Greenberg, editor), Ann. of Math. Studies, vol. 79, Princeton Univ. Press, Princeton, N. J., and Univ. of Tokyo Press, Tokyo, 1974, pp. 315–339.

[**170**] ——, *Schottky groups and circles*, Contributions to Analysis (Collection Dedicated to Lipman Bers; L. Ahlfors et al., editors), Academic Press, 1974, pp. 273–278.

[**171**] A. Marden and B. Maskit, *On the isomorphism theorem for Kleinian groups*, Invent. Math. **51** (1979), 9–14.

[**172**] G. A. Margulis, *Isometry of closed manifolds of constant negative curvature with the same fundamental group*, Dokl. Akad. Nauk SSSR **192** (1970), 736–737=Soviet Math. Dokl. **11** (1970), 722–723.

[**173**] ——, *Arithmetic properties of discrete subgroups*, Uspekhi Mat. Nauk **29** (1974), no. 1 (174), 49–98=Russian Math. Surveys **29** (1974), no. 1, 107–156.

[**174**] Olli Martio and Uri Srebro, *On the existence of automorphic quasimeromorphic mappings in* R^n, Ann. Acad Sci. Fenn. Ser. A I Math. **3** (1977), 123–130.

[**175**] ——, *Automorphic quasimeromorphic mappings in* R^n, Acta Math. **135** (1975), 221–247.

[**176**] Bernard Maskit, *A theorem on planar covering surfaces with applications to 3-manifolds*, Ann. of Math. (2) **81** (1965), 341–355.

[**177**] ——, *A characterization of Schottky groups*, J. Analyse Math. **19** (1967), 227–230.

[**178**] ——, *The conformal group of a plane domain*, Amer. J. Math. **90** (1968), 718–722.

[**179**] ——, *On a class of Kleinian groups*, Ann. Acad. Sci. Fenn. Ser. A I Math. No. 442 (1969).

[**180**] ——, *On boundaries of Teichmüller spaces and on Kleinian groups*. II, Ann. of Math. (2) **91** (1970), 607–639.

[**181**] ——, *On Klein's combination theorem*. III, Advances in the Theory of Riemann Surfaces (Proc. Conf., Stony Brook, N. Y., 1969; L. V. Ahlfors et al., editors), Ann. of Math. Studies, vol. 66, Princeton Univ. Press, Princeton, N. J., and Univ. of Tokyo Press, Tokyo, 1971, pp. 297–316.

[**182**] ——, *Self-maps on Kleinian groups*, Amer. J. Math. **93** (1971), 840–856.

[**183**] ——, *On Poincaré's theorem for fundamental polygons*, Advances in Math. **7** (1971), 219–230.

[**184**] ——, *Decompositions of certain Kleinian groups*, Acta Math. **130** (1973), 243–263.

[**185**] ——, *Intersections of component subgroups of Kleinian groups*, Discontinuous Groups and Riemann Surfaces (Proc. Conf., University Park, Md., 1973; L. Greenberg, editor), Ann. of Math. Studies, vol. 79, Princeton Univ. Press, Princeton, N. J., and Univ. of Tokyo Press, Tokyo, 1974, pp. 349–367.

[**186**] ——, *Uniformizations of Riemann surfaces*, Contributions to Analysis (Collection Dedicated to Lipman Bers; L. Ahlfors et al., editors), Academic Press, 1974, pp. 293–312.

[**187**] ——, *Moduli of marked Riemann surfaces*, Bull. Amer. Math. Soc. **80** (1974), 773–777.

[**188**] ——, *On the classification of Kleinian groups*. I, II, Acta Math. **135** (1975), 249–270; **138** (1977), 17–42.

[**189**] A. D. Mednykh, *On semidirect products of discontinuous transformation groups*, Dokl. Akad. Nauk SSSR **225** (1975), 1016–1017=Soviet Math. Dokl. **16** (1975), 1578–1580.

[**190**] ——, *On branched coverings of Riemann surfaces with the trivial group of covering transformations*, Dokl. Akad. Nauk SSSR **235** (1977), 1267–1269=Soviet Math. Dokl. **18** (1977), 1156–1158.

[**191**] ——, *On an example of a compact Riemann surface with trivial automorphism group*, Dokl. Akad. Nauk SSSR **237** (1977), 32–34=Soviet Math. Dokl. **18** (1977), 1396–1398.

[**192**] ——, *Determination of the number of nonequivalent coverings over a compact Riemann surface*, Dokl. Akad. Nauk SSSR **239** (1978), 269–271=Soviet Math. Dokl. **19** (1978), 318–320.

[**193**] ——, *On unramified coverings of compact Riemann surfaces*, Dokl. Akad. Nauk SSSR **244** (1979), 529–532=Soviet Math. Dokl. **20** (1979), 85–88.

[**194**] G. D. Mostow, *Strong rigidity of locally symmetric spaces*, Ann. of Math. Studies, vol. 78, Princeton Univ. Press, Princeton, N. J., and Univ. of Tokyo Press, Tokyo, 1973.

[**195**] David Mumford, *A remark on Mahler's compactness theorem*, Proc. Amer. Math. Soc. **28** (1971), 289–294.

[**196**] ——, *Stability of projective varieties*, L'Enseignement Math. (2) **23** (1977), 39–110; reprint, Monogr. Enseignement Math., no. 24, L'Enseignement Math., Geneva, 1977.([24])

[**197**] P. J. Myrberg, *Die Kapazität der singulären Menge der linearen Gruppen*, Ann. Acad. Sci. Fenn. Ser. A I Math.-Phys. No. 10 (1941).

[**198**] Zeev Nehari, *The Schwarzian derivative and schlicht functions*, Bull. Amer. Math. Soc. **55** (1949), 545–551.

[**199**] Rolf Nevanlinna, *Uniformisierung*, Springer-Verlag, 1953.

[**200**] S. P. Novikov, *Manifolds with free Abelian fundamental groups and their applications* (*Pontryagin classes, smoothness, multidimensional knots*), Izv. Akad. Nauk SSSR Ser. Mat. **30** (1966), 207–246; English transl., Amer. Math. Soc. Transl. (2) **71** (1968), 1–42.

[**201**] William F. Osgood, *A Jordan curve of positive area*, Trans. Amer. Math. Soc. **4** (1903), 107–112.

[**202**] C. D. Papakyriakopoulos, *On Dehn's lemma and the asphericity of knots*, Ann. of Math. (2) **66** (1957), 1–26.

[**203**] ——, *On solid tori*, Proc. London Math. Soc. (3) **7** (1957), 281–299.

[**204**] H. Poincaré, *Mémoire sur les groupes kleinéens*, Acta Math. **3** (1883/84), 49–92.

[**205**] ——, *Cinquième complément à l'Analysis situs*, Rend. Circ. Mat. Palermo **18** (1904), 45–110.

[**206**] ——, *Sur l'uniformisation des fonctions analytiques*, Acta Math. **31** (1908), 1–63.

[**207**] ——, *Oeuvres*. Vol. II, Gauthier-Villars, Paris, 1914.

[**208**] Christian Pommerenke, *On the Green's fundamental domain*, Math. Z. **156** (1977), 157–164.

([24])The authors cite a preprint.

[**209**] H. E. Rauch, *A transcendental view of the space of algebraic Riemann surfaces*, Bull. Amer. Math. Soc. **71** (1965), 1–39.

[**210**] Yu. G. Reshetnyak, *Representations in space with bounded distortion*, Internat. Congr. Math. (Nice, 1970): Papers of Soviet Mathematicians, "Nauka", Moscow, 1972, pp. 258–263. (Russian)

[**211**] Robert Riley, *Discrete parabolic representations of link groups*, Mathematika **22** (1975), 141–150.

[**212**] C. P. Rourke and B. J. Sanderson, *Introduction to piecewise-linear topology*, Springer-Verlag, 1972.

[**213**] H. L. Royden, *Automorphisms and isometries of Teichmüller spaces*, Advances in the Theory of Riemann Surfaces (Proc. Conf., Stony Brook, N. Y., 1969; L. A. Ahlfors et al., editors), Ann. of Math. Studies, vol. 66, Princeton Univ. Press, Princeton, N. J., and Univ. of Tokyo Press, Tokyo, 1971, pp. 369–383.

[**214**] V. V. Savin, *On moduli of Riemann surfaces*, Dokl. Akad. Nauk SSSR **196** (1971), 783–785=Soviet Math. Dokl. **12** (1971), 267–270.

[**215**] G. P. Scott, *Finitely generated* 3*-manifold groups are finitely presented*, J. London Math. Soc. (2) **6** (1972/73), 437–440.

[**216**] ——, *Compact submanifolds of* 3*-manifolds*, J. London Math. Soc. (2) **7** (1973/74), 246–250.

[**217**] H. Seifert and W. Threlfall, *Lehrbuch der Topologie*, Teubner, Leipzig, 1934.

[**218**] C. Weber and H. Seifert, *Die beiden Dodekaederräume*, Math. Z. **37** (1933), 237–253.

[**219**] Atle Selberg, *On discontinuous groups in higher-dimensional symmetric spaces*, Contributions to Function Theory (Internat. Colloq., Bombay, 1960), Tata Inst. Fund. Res., Bombay, 1960, pp. 147–164.

[**220**] V. G. Sheretov, *Invariant subspaces in a Teichmüller space*, Perm. Gos. Univ. Uchen. Zap. Mat. No. 103 (1963), 156–159. (Russian)

[**221**] Edwin H. Spanier, *Algebraic topology*, McGraw-Hill, 1966.

[**222**] George Springer *Introduction to Riemann surfaces*, Addison-Wesley, 1957.

[**223**] John R. Stallings, *On the loop theorem*, Ann. of Math. (2) **72** (1960), 12–19.

[**224**] ——, *On fibering certain* 3*-manifolds*, Topology of 3-Manifolds and Related Topics (Proc. Univ. of Georgia Inst., 1961; M. K. Fort, Jr., editor), Prentice-Hall, Englewood Cliffs, N. J., 1962, pp. 95–100.

[**225**] ——, *A topological proof of Grushko's theorem on free products*, Math. Z. **90** (1965), 1–8.

[**226**] Jacob Sturm and Meir Shinnar, *The maximal inscribed ball of a Fuchsian group*, Discontinuous Groups and Riemann Surfaces (Proc. Conf., University Park, Md., 1973; L. Greenberg, editor), Ann. of Math. Studies,

vol. 79, Princeton Univ. Press, Princeton, N. J., and Univ. of Tokyo Press, Tokyo, 1974, pp. 439–443.

[**227**] R. Sulanke and P. Wintgen, *Differentialgeometrie und Faserbündel*, Birkhäuser, 1972.

[**228**] Dennis Sullivan, *On the ergodic theory at infinity of an arbitrary discrete group of hyperbolic motions*, Riemann Surfaces and Related Topics: Proc. 1978 Stony Brook Conf. (I. Kra and B. Maskit, editors), Ann. of Math. Studies, vol. 97, Princeton Univ. Press, Princeton, N. J., and Univ. of Tokyo Press, Tokyo, 1981, pp. 465–496.

[**229**] Oswald Teichmüller, *Extremale quasikonforme Abbildungen und quadratische Differentiale*, Abh. Preuss. Akad. Wiss. Berlin Math.-Nat. Kl. **1939**, no. 22 (1940).

[**230**] ———, *Bestimmung der extremalen quasikonforme Abbildungen bei geschlossenen orientierten Riemannschen Flächen*, Abh. Preuss. Akad. Wiss. Berlin Math.-Nat. Kl. **1943**, no. 4.

[**231**] ———, *Veränderliche Riemannsche Flächen*, Deutsche Math. **7** (1942/44), 344–359.

[**232**] A. V. Tetenov, *Invariant components of Kleinian groups in space*, Materials Seventeenth All-Union Sci. Student Conf. Math., Novosibirsk. Gos. Univ., Novosibirsk, 1979, pp. 134–137. (Russian)

[**233**] ———, *Infinitely generated Kleinian groups in space*, Sibirsk. Mat. Zh. **21** (1980), no. 5, 88–99=Siberian Math. J. **21** (1980), 709–717.

[**234**] William Thurston, *The geometry and topology of 3-manifolds*, Lecture notes, Princeton Univ., Princeton, N. J., 1978.

[**235**] È. B. Vinberg, *Discrete groups generated by reflections in Lobachevsky spaces*, Mat. Sb. **72**(**114**) (1967), 471–488=Math. USSR Sb. **1** (1967), 429–444.

[**236**] ———, *Some examples of crystallographic groups in Lobachevsky spaces*, Mat. Sb. **78**(**120**) (1969), 633–639=Math. USSR Sb. **7** (1969), 617–622.

[**237**] È. B. Vinberg and O. V. Shvartsman, *Riemann surfaces*, Itogi Nauki: Algebra, Topologiya, Geometriya, vol. 16, VINITI, Moscow, 1978, pp. 191–245=J. Soviet Math. **14** (1980), no. 1, 985–1020.

[**238**] Friedhelm Waldhausen, *Eine Verallgemeinerung des Schleifensatzes*, Topology **6** (1967), 501–504.

[**239**] ———, *Gruppen mit Zentrum und 3-dimensionale Mannigfaltigkeiten*, Topology **6** (1967), 505–517.

[**240**] ———, *On irreducible 3-manifolds which are sufficiently large*, Ann. of Math. (2)**87** (1968), 56–88.

[**241**] Hermann V. Waldinger, *On the subgroups of the Picard group*, Proc. Amer. Math. Soc. **16** (1965), 1373–1378.

[**242**] André Weil, *Modules des surfaces de Riemann*, Sém. Bourbaki, 1957/58, Exposé 168, Secrétariat Math., Paris, 1958; reprint, Benjamin, 1966.

[**243**] J. H. C. Whitehead, *On 2-spheres in 3-manifolds*, Bull. Amer. Math. Soc. **64** (1958), 161–166.

[**244**] ——, *On finite cocycles and the sphere theorem*, Colloq. Math. **6** (1958), 271–281.

[**245**] Norbert J. Wielenberg, *Discrete Moebius groups: fundamental polyhedra and convergence*, Amer. J. Math. **99** (1977), 861–877.

[**246**] ——, *The structure of certain subgroups of the Picard group*, Math. Proc. Cambridge Philos. Soc. **84** (1978), 427–436.

[**247**] A. Wiman, *Über die algebraischen Curven von den Geschlechtern $p = 4, 5$ und 6, welche eindeutige Transformationen in sich besitzen*, Bichang Kongl. Svenska Vetenskaps-Akad. Handlingar **21** (1895/96), afd. I, no. 3.

[**248**] Scott Wolpert, *The finite Weil-Petersson diameter of Riemann space*, Pacific J. Math. **70** (1977), 281–288.

[**249**] Robert Zarrow, *Classical and non-classical Schottky groups*, Duke Math. J. **42** (1975), 717–724.

[**250**] I. V. Zhuravlev, *Univalent functions with quasiconformal extensions, and Teichmüller spaces*, Preprint, Inst. Mat. Sibirsk. Otdel. Akad. Nauk SSSR, Novosibirsk, 1979. (Russian)

Supplementary Bibliography

[**251**] William Abikoff, *The real analytic theory of Teichmüller space*, Lecture Notes in Math., vol. 820, Springer-Verlag, 1980.

[**252**] D. V. Anosov, *Geodesic flows on closed Riemannian manifolds of negative curvature*, Trudy Mat. Inst. Steklov. **90** (1967)=Proc. Steklov Inst. Math. **90** (1967).

[**253**] B. N. Apanasov, *Geometrically finite groups of transformations of space*, Sibirsk. Mat. Zh. **23** (1982), no. 6, 16–27=Siberian Math. J. **23** (1982), 771–780.

[**254**] ——, *Geometrically finite hyperbolic structures on manifolds*, Ann. Global Anal. and Geom. **1** (1983), no. 3, 1–22.

[**255**] ——, *Condition of conformal rigidity of hyperbolic manifolds with boundaries*, Analytic Functions (Błażejewko, 1982), Lecture Notes in Math., vol. 1039, Springer-Verlag, 1983, pp. 1–8.

[**256**] ——, *Discrete transformation groups and structures on manifolds*, "Nauka", Novosibirsk, 1983. (Russian)

[**257**] ——, *Incomplete hyperbolic manifolds and local finiteness of a tesselation of space by polyhedra*, Dokl. Akad. Nauk SSSR **273** (1983), 777–781=Soviet Math. Dokl. **28** (1983), 686–690.

[**258**] ——, *Parabolic vertices and finiteness properties for Kleinian groups in n-dimensional space*, Sibirsk. Mat. Zh. **25** (1984), no. 4, 9–27=Siberian Math. J. **25** (1984), 516–530.

[**259**] ——, *Cusp-neighbourhoods for Kleinian groups in space*, Complex Analysis and Its Applications to Partial Differential Equations (KTB, Wiss. Beiträge 1984/58 (M 35)), Martin-Luther-Univ., Halle (S), 1984, pp. 37–39.

[**260**] ——, *Cusp ends of hyperbolic manifolds*, Ann. Global Anal. and Geom. **3** (1985), no. 1, 1–12.

[**261**] ——, *Tesselations of space by polyhedra and deformations of incomplete hyperbolic structures*, Sibirsk. Mat. Zh. **27** (1985), no. 4, 9–27; English transl. in Siberian Math. J. **27** (1985).

[**262**] ——, *Deformations of conformal structures on manifolds*, Abstracts Internat. Conf. Complex Anal. and Appl., Varna, 1985, p. 37.

[**263**] T. M. Bandman, *Surjective holomorphic mappings of projective manifolds*, Sibirsk. Mat. Zh. **22** (1981), no. 2, 48–56=Siberian Math. J. **22** (1981), 204–210.

[**264**] Lipman Bers, *Nielsen extensions of Riemann surfaces*, Ann. Acad. Sci. Fenn. Ser. A I Math. **2** (1976), 29–34.

[**265**] ——, *An extremal problem for quasiconformal mappings and a theorem by Thurston*, Acta Math. **141** (1978), 73–98.

[**266**] ——, *On iterates of hyperbolic transformations of Teichmüller space*, Amer. J. Math. **105** (1983), 1–11.

[**267**] Igor Dolgachev, *Integral quadratic forms: applications to algebraic geometry (after V. Nikulin)*, Sém. Bourbaki 1982/83, Exposé 611, Astérisque No. 105–106, Soc. Math. France, Paris, 1983, pp. 251–278.

[**268**] Clifford J. Earle, *On the Carathéodory metric in Teichmüller spaces*, Discontinuous Groups and Riemann Surfaces, (Proc. Conf., University Park, Md., 1973; L. Greenberg, editor), Ann. of Math. Studies, vol. 79, Princeton Univ. Press, Princeton, N. J., and Univ. of Tokyo Press, Tokyo, 1974, pp. 99–103.

[**269**] A. Fathi et al., *Travaux de Thurston sur les surfaces* (Sém. Orsay, 1976/77), Astérisque No. 66–67, Soc. Math. France, Paris, 1979.

[**270**] Jane Gilman, *On the Nielsen type and the classification for the mapping class group*, Advances in Math. **40** (1981), 68–96.

[**271**] William M. Goldman, *Conformally flat manifolds with nilpotent holonomy and the uniformization problem for 3-manifolds*, Trans. Amer. Math. Soc. **278** (1983), 573–583.

[**272**] N. A. Gusevskiĭ, *On doubly generated Kleinian groups*, Sibirsk. Mat. Zh. **23** (1982), no. 1, 183–186. (Russian)

[**273**] ——, *On the fundamental group of a manifold of negative curvature*, Dokl. Akad. Nauk SSSR **268** (1983), 777–781=Soviet Math. Dokl. **27** (1983), 140–144.

[**274**] ——, *Schottky groups in space*, Dokl. Akad. Nauk SSSR **280** (1985), 30–33=Soviet Math. Dokl. **31** (1985), 21–24.

[**275**] ——, *Uniformization of conformal structures on three-dimensional manifolds*, Abstracts, Internat. Conf. Complex Analy. and Appl., Varna, 1985, p. 80.

[**276**] Alan Howard and Andrew J. Sommese, *On the theorem of de Franchis*, Ann. Scuola Norm. Sup. Pisa Cl. Sci. (4) **10** (1983), 429–436.

[**277**] Irwin Kra, *Deformations of Fuchsian groups*, Duke Math. J. **36** (1969), 537–546.

[**278**] ——, *On the Nielsen-Thurston-Bers type of some self-maps of Riemann surfaces*, Acta Math. **146** (1981), 231–270.

[**279**] ——, *On the vanishing of and spanning sets for Poincaré series for cusp forms*, Acta Math. **153** (1984), 47–116.

[**280**] Irwin Kra and Bernard Maskit, *Remarks on projective structures*, Riemann Surfaces and Related Topics: Proc. 1978 Stony Brook Conf. (I. Kra and B. Maskit, editors), Ann. of Math. Studies, vol. 97, Princeton Univ. Press, Princeton, N.J., and Univ. of Tokyo Press, Tokyo, 1981, pp. 343–360.

[**281**] S. L. Krushkal′, *Invariant metrics in Teichmüller spaces*, Sibirsk. Mat. Zh. **22** (1981), no. 2, 209–212. (Russian)

[**282**] ——, *Invariant metrics on spaces of closed Riemann surfaces*, Sibirsk. Mat. Zh. **26** (1985), no. 2, 108–114=Siberian Math. J. **26** (1985) (to appear).

[**283**] N. H. Kuiper, *Compact spaces with a local structure determined by the group of similarity transformations in* E^n, Nederl. Akad. Wetensch. Proc. **53** (1950), 1178–1185=Indag. Math. **12** (1950), 411–418.

[**284**] Folke Lannér, *On complexes with transitive groups of automorphisms*, Medd. Lunds Univ. Mat. Sem. **11** (1950).

[**285**] Wilhelm Magnus, *Noneuclidean tesselations and their groups*, Academic Press, 1974.

[**286**] A. I. Mal′tsev, *On the faithful representation of infinite groups by matrices*, Mat. Sb. **8(50)** (1940), 405–422; English transl., Amer. Math. Soc. Transl. (2) **45** (1965), 1–18.

[**287**] W. Mangler, *Die Klassen von topologischen Abbildungen einer geschlossenen Fläche auf sich*, Math. Z. **44** (1938/39), 541–554.

[**288**] G. A. Margulis, *Discrete groups of motions of manifolds of nonpositive curvature*, Proc. Internat. Congr. Math. (Vancouver, 1974), Vol. 2, Canad. Math. Congr., Montréal, 1975, pp. 21–34; English transl. in Amer. Math. Soc. Transl. (2) **109** (1977).

[**289**] ——, *The arithmeticity of irreducible lattices in semisimple groups of rank greater than* 1, supplement to the Russian transl. of M. S. Raghunathan, *Discrete subgroups of Lie groups*, "Mir", Moscow, 1977, pp. 277–313. (Russian)

[**290**] A. D. Mednykh, *On the Hurwitz problem on the number of nonequivalent coverings over a compact Riemann surface*, Sibirsk. Mat. Zh. **23** (1982), no. 3, 155–160=Siberian Math. J. **23** (1982), 415–420.

[**291**] ——, *On the solution of the Hurwitz problem on the number of nonequivalent coverings over a compact Riemann surface*, Dokl. Akad. Nauk SSSR **261** (1981), 537–542=Soviet Math. Dokl. **24** (1981), 541–545.

[**292**] ——, *Nonequivalent coverings of Riemann surfaces with a given type of branching*, Sibirsk. Mat. Zh. **25** (1984), no. 4, 120–142=Siberian Math. J. **25** (1984), 606–625.

[**293**] ——, *On the number of subgroups of the fundamental group of a closed nonnegative surface*, Seventeenth All-Union Algebra Conf., Abstracts of Reports, Part 1, Inst. Mat. Akad. Nauk Beloruss. SSR, Minsk, 1983, p. 125. (Russian)

[**294**] Jakob Nielsen, *Untersuchungen zur Topologie der geschlossenen zweiseitigen Flächen.* I, II, Acta Math. **50** (1927), 189–358; **53** (1929), 1–76.

[**295**] V. V. Nikulin, *On arithmetic groups generated by reflections in Lobachevsky spaces*, Izv. Akad. Nauk SSSR Ser. Mat. **44** (1980), 637–669=Math. USSR Izv. **16** (1981), 573–601.

[**296**] ——, *On the classification of arithmetic groups generated by reflections in Lobachevsky spaces*, Izv. Akad. Nauk SSSR Ser. Mat. **45** (1981), 113–142=Math. USSR Izv. **18** (1982), 99–123.

[**297**] David B. Patterson, *The fundamental group of the modulus space*, Michigan Math. J. **26** (1979), 213–223.

[**298**] Atle Selberg, *Recent developments in the theory of discontinuous groups of motions of symmetric spaces*, Proc. Fifteenth Scandinavian Congr. (Oslo, 1968), Lecture Notes in Math., vol. 118, Springer-Verlag, 1970, pp. 99–120.

[**299**] Dennis Sullivan, *Growth of positive harmonic functions and Kleinian group limit sets of zero planar measure and Hausdorff dimension two*, Geometry Sympos. (Utrecht, 1980), Lecture Notes in Math., vol. 894, Springer-Verlag, 1981, pp. 127–144.

[**300**] ——, *Discrete conformal groups and measurable dynamics*, Bull. (New Ser.) Amer. Math. Soc. **6** (1982), 57–74.

[**301**] ——, *Quasiconformal homeomorphisms and dynamics.* II: *Structural stability implies hyperbolicity for Kleinian groups*, preprint, Inst. Hautes Études Sci., Bures-sur-Yvette, 1982.

[**302**] William P. Thurston, *On the geometry and dynamics of diffeomorphisms of surfaces*, preprint.

[**303**] ——, *Three dimensional manifolds, Kleinian groups and hyperbolic geometry*, Bull. (New Ser.) Amer. Math. Soc. **6** (1982), 357–381.

[**304**] Pukka Tukia, *The Hausdorff dimension of the limit set of a geometrically finite Kleinian group*, Acta Math. **152** (1984), 127–140.

[**305**] ——, *Quasiconformal extension of quasisymmetric mappings compatible with a Möbius group*, Acta Math. **154** (1985), 153–193.

[**306**] È. B. Vinberg, *On groups of units of certain quadratic forms*, Mat. Sb. **87(129)** (1972), 18–36=Math. USSR Sb. **16** (1972), 17–35.

[**307**] ——, *On unimodular integral quadratic forms*, Funktsional. Anal. i Prilozhen. **6** (1972), no. 2, 24–31=Functional Anal. Appl. **6** (1972), 105–111.

[**308**] ——, *The absence of crystallographic groups of reflections in Lobachevsky spaces of large dimension*, Funktsional. Anal. i Prilozhen. **15** (1981), no. 2, 67–68=Functional Anal. Appl. **15** (1981), 128–130.

[**309**] ——, *The absence of crystallographic groups of reflections in Lobachevsky spaces of large dimension*, Trudy Moskov. Mat. Obshch. **47** (1984), 68–102=Trans. Moscow Math. Soc. **1985**, no. 1(47), 75–112.

[**310**] È. B. Vinberg and I. M. Kaplinskaya, *On the groups* $O_{18,1}(\mathbf{Z})$ *and* $O_{19,1}(\mathbf{Z})$, Dokl. Akad. Nauk SSSR **238** (1978), 1273–1275 Soviet Math. Dokl. **19** (1978), 194–197.

Subject Index

ABCDEFGHIJ–AMS/AP–89876